S0-AIY-016

RAINFORESTS OF THE WORLD

A Reference Handbook

Second Edition

Other Titles in ABC-CLIO's
CONTEMPORARY
WORLD ISSUES
Series

World Population, Geoffrey Gilbert
Biodiversity, Anne Becher
Wilderness Preservation, Kenneth A. Rosenberg
The Ozone Dilemma, David E. Newton
Tobacco, Harold V. Cordry
Recycling in America, 2d Edition, Debra L. Strong
Animals and the Law, Jordan Curnutt
Natural Disasters: Floods, E. Willard Miller & Ruby M. Miller
Agricultural Crisis in America, Dana L. Hoag

Books in the Contemporary World Issues series address vital issues in today's society such as terrorism, sexual harassment, homelessness, AIDS, gambling, animal rights, and air pollution. Written by professional writers, scholars, and nonacademic experts, these books are authoritative, clearly written, up-to-date, and objective. They provide a good starting point for research by high school and college students, scholars, and general readers, as well as by legislators, businesspeople, activists, and others.

Each book, carefully organized and easy to use, contains an overview of the subject; a detailed chronology; biographical sketches; facts and data and/or documents and other primary-source material; a directory of organizations and agencies; annotated lists of print and nonprint resources; a glossary; and an index.

Readers of books in the Contemporary World Issues series will find the information they need in order to better understand the social, political, environmental, and economic issues facing the world today.

DISCARDED
New Hanover County Public Library
from

RAINFORESTS OF THE WORLD

A Reference Handbook

Second Edition

Kathlyn Gay

NEW HANOVER COUNTY
PUBLIC LIBRARY
201 CHESTNUT STREET
WILMINGTON, NC 28401

CONTEMPORARY
WORLD ISSUES

A B C ☰ C L I O

Santa Barbara, California   Denver, Colorado   Oxford, England

Copyright © 2001 by Kathlyn Gay

All rights reserved. No part of this publication may be reproduced, stored in a retrieval system, or transmitted, in any form or by any means, electronic, mechanical, photocopying, recording, or otherwise, except for the inclusion of brief quotations in a review, without prior permission in writing from the publishers.

Library of Congress Cataloging-in-Publication Data

Gay, Kathlyn.
    Rainforests of the world : a reference handbook / Kathlyn Gay.— 2nd ed.
        p. cm.
Includes bibliographical references and index (p.   ).
    ISBN 1-57607-424-2 (hardcover : alk. paper)
    1. Rain forest ecology. 2. Rain forests. 3. Rain forest conservation. 4. Deforestation—Environmental aspects—Tropics.
I. Title.
    QH541.5.R27 G39 2001
    577.34—dc21   2001002780

This book is also available on the World Wide Web as an e-book. Visit abc-clio.com for details.

06 05 04 03 02 01   10 9 8 7 6 5 4 3 2 1

ABC-CLIO, Inc.
130 Cremona Drive, P.O. Box 1911
Santa Barbara, California 93116-1911

This book is printed on acid-free paper ⊚
Manufactured in the United States of America

Contents

Preface

Many people around the world are aware that tropical rainforests are being destroyed at a rapid rate; indeed, 80 to 90 acres (40 hectares) of tropical rainforest are destroyed every minute. An area of tropical forest the size of New York State disappears each year. But not so well known is the major loss of rainforests in the temperate regions of the world, including western North America, New Zealand, Tasmania, Chile, Argentina, portions of Japan, northwestern Europe, and the Black Sea coasts of Turkey and Georgia.

Both temperate and tropical rainforests are vital parts of the Earth's life-support systems, supplying oxygen, regulating climate, generating precipitation, conserving soil, and providing ecosystems for a multitude of diverse plant and animal species and homelands for thousands of indigenous people—those native to the land.

Over the past few decades an increasing number of botanists, ecologists, climatologists, and other scientists have expressed their concerns about the destruction of all types of forests worldwide. Why are rainforests being obliterated? What benefits do rainforests provide to human and animal populations around the world? What are the effects of rainforest destruction? What can be done to preserve the rainforests?

These questions, basic to the research being conducted by those attempting to preserve rainforests, introduce just a few of the issues discussed in Chapter 1, which is an overview of rainforests worldwide. Problems and possible solutions to rainforest destruction have created controversies on a global as well as a local scale, and these are covered in Chapter 2. An effort to conserve forests began in the early 1800s in the United States, and the conservation and preservation of rainforests in particular has taken on increasing intensity since the 1990s, as the chronology in Chapter 3 indicates.

Sketches of individuals and events associated with rainforests are the subject of Chapter 4, and Chapter 5 provides charts, tables, graphics, and documents on rainforest destruction and the effects on human, plant, and animal life. Groups working to protect and conserve rainforests and to replant destroyed areas and groups opposed to taking private property to protect rainforests and bio-diversity are described in Chapter 6. Addresses, including Web addresses, and phone numbers are also part of Chapter 6.

Chapter 7 includes annotated bibliographic references on rainforests; rainforest animals, plants, and products; people who live in and depend on the rainforests for survival; the effects of rainforest destruction; and efforts to save these valuable natural resources. Nonprint references are also listed and annotated. A glossary includes definitions of some of the terms associated with rainforests around the world.

A special thank you to Karen Hamilton for her help with research and chart preparation for this edition of *Rainforests of the World*.

1

Introduction

Rainforest Locations

Forest experts have classified dozens of different types of forests around the world, including several categories for rainforests, based on such factors as location, soil type, climate, and the extent to which the forests have remained in their "natural" state, undisturbed by human activities. For example, some rainforests cover lowland areas, others grow in mountainous regions, and still others spread across plains that are covered with floodwaters at certain times each year.

The two main types of rainforests discussed throughout this book are tropical and temperate rainforests. As the terms suggest, these forests are located in either the tropical or temperate zones of the world.

Tropical Rainforests

Most tropical rainforests fit into one of three basic types, each defined by the amount of precipitation: a tropical dry forest, a tropical moist forest, or a true tropical rainforest. True tropical rainforests are closest to the equator and receive more precipitation than the other two types. In fact, calling a tropical dry forest a rainforest may seem a misnomer because rain may not fall in these locations for months at a time. Generally, a true rainforest receives about 80 to 120 inches (203 to 305 centimeters) of precipitation or more (sometimes much more) per year.

Tropical rainforests make up 6 to 7 percent of the earth's surface and are located in a band covering 3.4 million square miles and extending 10 degrees in both directions from the equator, to south of the Tropic of Cancer and north of the Tropic of Capricorn. This band of tropical rainforests includes parts of Central and South America, Africa, Asia, and the United States. Year-round, the temperature in these areas averages 80 degrees Fahrenheit (27 degrees Celsius).

Most tropical forests are in the warm, wet areas 10 degrees on either side of the equator, where they flourish best. Year-around the temperature in these areas averages 80 degrees Fahrenheit (27 degrees Celsius). Perhaps the best known of all tropical forests is in Amazonia—the Amazon River basin. Covering an area almost the size of the United States, this forest region stretches from the foothills of the Andes Mountains to the Atlantic seacoast in Brazil. Other tropical rainforests include those in Papua New Guinea, the Congo Basin and the Democratic Republic of the Congo in Africa, the islands of Madagascar, the Malaysian Peninsula, northern Thailand, southeastern Mexico, and parts of Colombia and Ecuador. Tropical rainforests under U.S. jurisdiction include Wao Kele O Puna on the island of Hawaii, the Caribbean National Forest in Puerto Rico, and rainforests on the three small islands of American Samoa.

Temperate Rainforests

Nearly all of the original rainforests in temperate zones were destroyed long ago, and because of varying classification methods and the continuing massive deforestation of temperate forests worldwide, there is limited information on the remaining stands of temperate rainforest. But coastal rainforests in temperate regions are relatively healthy, and some ecologists have attempted to assess the distribution and status of coastal temperate rainforests.

Two conservation organizations, Ecotrust and Conservation International, supported studies of coastal temperate rainforests and in 1992 reported that an estimated 80–100 million acres (30–40 million hectares) of coastal temperate rainforest once existed. Today less than half of that remains, although no one is sure of the total. The forest areas that still exist are primarily on the western edges of continents and cover coastal regions in North and South America, New Zealand, Tasmania, portions of Japan, northwestern Europe, Turkey, and Georgia (part of the former Soviet Union) (Ecotrust and Conservation International 1992, 3).

The North American coastal rainforests are perhaps the most studied and inventoried. Accounting for about 40 to 50 percent of the world's remaining coastal temperate rainforests, they are found along a narrow, fifteen-mile-wide strip of coastal land between the sea and the coastal mountain ranges. Because of the interaction of the oceans and mountains, there is high rainfall, with at least 80 inches (203 centimeters) of annual precipitation (rain, snow, or fog). Some of these forests receive up to 200 inches (508 centimeters) of precipitation per year.

The temperate rainforests of the Pacific Northwest stretch from southern Oregon to southeastern Alaska, including coastal sections of British Columbia (B.C.) in Canada. The Olympic National Park in Washington, the Carmanah Valley on Vancouver Island, B.C., and Alaska's two largest national forests, the Chugach and the Tongass, are among the temperate rainforests of North America. Alaskan rainforests are along a thousand-mile arc on the Pacific coast between the communities of Ketchikan and Kodiak and include 22.5 million acres (9.1 million hectares) of ancient forest, where giant trees, hundreds of feet tall, live up to a thousand years.

In the southern hemisphere, the largest remaining area of coastal temperate rainforest is in Chile. Ecologists who have analyzed Chilean rainforests conclude that these forests "are the most species-rich in the world, due to a complex landscape that has produced diverse habitats and species composition." Among the species found here is the alerce cedar, the largest conifer in South America. The alerce may be related to the giant Californian sequoias, and like these northern neighbors it has a very long life span—up to four thousand years (World Rainforest Movement 1992, 6).

Vital Support Systems

Since the 1960s and 1970s, an increasing number of botanists, ecologists, climatologists, and other scientists have been studying rainforests, which are vital parts of the Earth's life-support systems. Rainforests generate precipitation, conserve soil, supply oxygen, store carbon, regulate climate, and provide ecosystems for a multitude of diverse species of plants and animals.

Wherever they are located, forests play a major role in generating rain by returning moisture to the atmosphere in a process known as *evapotranspiration*. That is, water from trees and plants growing in the forest evaporates and becomes part of the hydrologic, or water, cycle. About half the rainfall over Amazonia returns to the atmosphere through evapotranspiration. Moisture in the atmosphere from this process affects regional and global climate. In the major tropical atmospheric circulation pattern known as Hadley cell circulation, moist air rises, forms clouds, and sheds moisture as rain.

Forests in both temperate and tropical zones have elaborate root systems that hold soil in place, preventing erosion. The forest root system also absorbs rainfall, helping to regulate water runoff. Water is stored and slowly released throughout the year, replenishing groundwater supplies and maintaining the flow of water in rivers and streams.

Vegetation, forest litter, and living organisms hold nutrients that are essential to the growth of forest plants and trees. Decaying matter releases nutrients into the soil, and some tree leaves and branches also obtain nutrients from rain as it falls or from plants and algae that grow on trees. When trees are cut, the nutrients that remain in the soil soon wash out. After logging, fairly rapid regrowth may take place. However, soil erosion may prevent full recovery, and the biological diversity of a logged area will not equal that of the original forest.

Forests are crucial in the cycling of six basic elements: carbon, oxygen, nitrogen, hydrogen, phosphorous, and sulfur, which make up 95 percent of the world's living matter. Since there is a fixed supply of these elements, life on earth depends on their efficient cycling through the soil, the hydrosphere (water and water vapor), lake and ocean sediments, flora, fauna, and the atmosphere. This cycling is accomplished in a variety of ways. For example, plant-eating animals deposit nutrients in their feces, and rainwater washes nutrients from one location to another through the soil. Another example is *mycorrhizae*—the relationship between fungi and the roots of trees; the fungi are nourished by the tree roots and take up nutrients from the soil to benefit their host.

Carbon, one of the most important of the chemical elements cycled in rainforests, usually combines with other elements to form what seems to be an unlimited variety of compounds. In fact, all living things are made up of carbon compounds. Green plants absorb carbon dioxide (CO_2) from the atmosphere in a process known as *carbon fixation:* The green pigment in plants, chlorophyll, captures sunlight and converts it to chemical energy in the process of photosynthesis. Next, this chemical energy is used to combine CO_2 and water, producing simple carbohydrates called glucose, an important source of energy for animals, including humans. During photosynthesis, plants also release the oxygen needed for survival. In turn, when people and other animals oxidize, or use up, food, they exhale CO_2. Because of this continued cycle, the supply of CO_2 and oxygen remains fairly stable.

Carbon is stored in various reservoirs such as the atmosphere and the oceans, the soil, and living plants and animals, and also in fossilized forms such as oil, gas, and coal deposits. Driven by physical and biological forces, carbon circulates naturally among these holding places. Forests are some of the major reservoirs for carbon. But the amount of carbon held in the various reservoirs is changing because of human activity.

Cutting down vast numbers of trees, whether in temperate or tropical forests, releases CO_2 into the atmosphere, adding billions of tons to the buildup of CO_2 from such activities as burning fossil fuels for energy, manufacturing, and transportation. CO_2 and other gases have accumulated in the atmosphere, trapping heat that reflects from the earth. Many scientists believe that this is causing an overall rise in the temperature of the planet, known as the *greenhouse effect* or *global warming*. However, global warming is a controversial concept still and scientists debate whether the greenhouse effect could alter rainfall, wind, and heat patterns around the world.

Rainforest Habitats

Many people think of rainforests as dark, intimidating "jungles," as they are so often depicted in movies. But "jungle" is not a synonym for "rainforest," and a rainforest is much more than dense growth. Those who explore and study rainforests use phrases like "symphony of life," "plush vegetation," "magical scents," and "rich eternal rhythms" to describe rainforest habitats.

Tropical Habitats

Because of their climate, tropical forests contain more diverse plant and animal life than any other place on Earth. At least 50 percent and perhaps up to 90 percent of all living species can be found in tropical forests. An estimated 50 million animal species, including millions of insect species, flourish in tropical rainforest areas. No one knows exactly how many species the rainforests support because scientists have only recently begun to classify and inventory species in various rainforest locations. Mammals that make the rainforest their home include various types of cats, deer, wild pigs, monkeys, sloths, marsupials, rats, and mice. An effort initiated in 1989 to categorize species in Costa Rican rainforests resulted in estimates of 500,000 different animal and plant species, among them an estimated 365,000 insect, spider, and tick species. According to the National Academy of Sciences, insects in the rainforests are so abundant that a single tree in the rainforests of Panama may be home to more than 1,700 species. In the rainforest of Sarawak, Malaysia, biologists have identified 3,000 species of butterfly and moth.

Life in a tropical rainforest exists in layers of vegetation that form separate habitats, beginning at the top, about 200 feet (60 meters) high, and going down all the way to the forest floor. At the very top, the upper canopy, a mass of treetops intertwines with vines and colorful, flowering plants. This dense top tier of the forest also supports a diverse number of mammals, birds, and insects. The upper canopy is home for the Guiana crested eagle and the harpy eagle (both endangered). Howler monkeys, fruit bats, sloths, and many other mammals and numerous birds, such as parrots, spend their lives in the upper canopy. At the next level, the lower canopy, birds, butterflies, spiders, iguanas, and other animals forage along the branches.

The next layer of growth, called the *understory*, begins about 50 to 60 feet (15 to 18 meters) above the ground, rising to a little more than 80 feet (25 meters). Since this tier is cooler and more humid than the canopy, fewer plants grow here. Finally, there is the forest floor. Although it is usually bare except for fallen leaves, decaying plants, and some sprouting plants, this part of the forest is, in effect, a recycling center. Fungal enzymes help break down fallen leaves, and an army of ants, beetles, termites, worms, and many other organisms as well as foraging birds clean up the forest floor. At the same time, these creatures help renew the forest by planting seeds or providing nutrients for continued growth.

Temperate Habitats

A mass of living greenery carpets the floor of a temperate rainforest. Flowers, ferns, and fungi abound. Some of the large mammals in the rainforest include elk, deer, mountain lion, black bear, beaver, and raccoon. Fish are plentiful in rainforest streams, and a variety of birds make their homes in the trees and shrubs.

In the protected temperate rainforests of the Quinault, Queets, and Hoh Rivers in the Pacific Northwest are found "some of the most spectacular examples of undisturbed Sitka

spruce/western hemlock forests in the lower forty-eight states,"
according to the U.S. National Park Service (NPS 2000). Because
mountains to the east protect the coastal areas from severe
weather extremes, the rainforest temperature seldom drops
below freezing and summertime highs rarely exceed 80 degrees
Fahrenheit. Along with Sitka spruce and western hemlock, some
of which reach 300 feet in height and 23 feet in circumference,
Douglas fir, western red cedar, bigleaf maple, red alder, vine
maple, and black cottonwood are also found throughout the rainforests. As the National Park Service notes:

> Nearly every bit of space is taken up with a living
> plant. Some plants even live on others. These are the epi-
> phytes, plants that do not come into contact with the
> earth, but also are not parasites. They are partly respon-
> sible for giving the rainforest its "jungly" appearance.
>
> Mosses, lichens and ferns cover just about anything
> else. . . . But because of this dense ground cover it is
> hard for seedlings to get a start. Many seedlings germi-
> nate on fallen, decaying trees. As they grow they send
> their roots down the log to the ground. Eventually the
> log rots completely away and a row of young trees is left,
> up on stilt-like roots, all in a row. The thick and protec-
> tive vegetation also provides excellent habitats for the
> animals of the rain forest. In turn, they contribute to the
> health of the forest by keeping the rampant vegetation
> under control by browsing.

The stands of ancient forests, sometimes called old-growth
forests, within the temperate rainforests of the Pacific Northwest
are a complex interdependent system of living things. All parts of
the system are interrelated, and each organism depends on another part of the system for survival. If one part of the system
dies or is damaged, the entire web of life is affected. For example,
when lumber companies log in an old-growth forest, they frequently *clear-cut*, or cut all the trees in hundreds of patches of
about 40 to 60 acres (16 to 24 hectares) each. They also clear off
land to build roads leading into the logging areas, a practice that
has severely fragmented forest habitats.

Products from Tropical Forests

Hundreds of items that people in industrialized nations use on a
daily basis come from tropical rainforests. They include food-

stuffs (fruits, vegetables, nuts, spices, teas, and ingredients for soft drinks), houseplants, fibers, rubber products, building materials, oils, and ingredients for perfumes, cosmetics, and pharmaceuticals. For example, thaumatin, a compound derived from the katemfe bush of the West African rainforest, has been called the sweetest substance in the world—it is 100,000 times sweeter than table sugar and could be processed for commercial use as a sugar substitute. Another West African plant, the Calabar bean, is the source of compounds for some insecticides.

A number of rainforest plants and trees produce oil that can be used as fuel. Some nut trees produce a crude oil that can substitute for gasoline. Sap from Amazonian copaiba trees has the potential to become a substitute fuel for motor vehicles. Its properties are identical to diesel fuel and can be used to power a truck.

Some of the most important rainforest products are medicinal drugs. In rainforest areas, between 80 and 90 percent of indigenous populations—those native to the land—depend on plants for medicinal purposes. Nearly one-fourth of all pharmaceuticals that Americans use originally derived from tropical plants, which provide the chemicals for processing synthetic drugs. Quinine, for example, was developed from the bark of the cinchona tree and is used to treat malaria and pneumonia. Diosgenin, a primary ingredient in cortisone, is produced from rainforest plants that thrive in Belize, Guatemala, and Mexico. Curare, a plant extract from the Amazon, is an ingredient for muscle relaxants. Drugs that stimulate the heart and respiratory system, anesthetics, tumor inhibitors, contraceptives, and anticancer medicines come from tropical rainforests. The Black Bean (*Castanospermum australe*) from the Queensland, Australia, rainforest is one of the world's great hopes for an AIDS cure.

Scientists have identified more than 2,000 tropical plants that contain substances for treating various types of cancer, and some say the total could go as high as 3,000 plants. One example is the rosy periwinkle, found in the rainforest of Madagascar Island, off the southeast coast of Africa. Periwinkle's tiny pink flower is the source of two compounds: one for the treatment of childhood leukemia and another for the treatment of Hodgkin's disease. Madagascar's eastern forest, in fact, is said to be a "biotic treasure house" because 12,000 known plant species and 190,000 known animal species originated there. But so far at least half of the original plants and animals have been lost along with a vast portion of the forest—only about 10 percent of the original forest remains.

Although scientists long considered plant and herbal treatments as a type of magic or "faith healing," that view began to change in the late 1980s and early 1990s. As forests have disappeared worldwide, more and more research efforts have been initiated to investigate the chemical properties of tropical plants. For example, the National Cancer Institute, headquartered in Maryland, has sent *ethnobotanists,* scientists who study the interrelationships between plants and people, to various countries to find plants that Native healers use for medicinal purposes. The institute and its researchers are concerned that not only the plants but also the healers will disappear before their curative secrets are revealed. For researchers, losing plants and knowledge about their chemical structures is akin to losing the great libraries of Alexandria in ancient Greece.

Currently more than 200 pharmaceutical companies worldwide and some agencies of the U.S. government are sponsoring projects to study rainforest plants used by indigenous healers and shamans (spiritual leaders) for the survival, health, and well-being of their communities. Many researchers contend that rainforest plants represent a gold mine of beneficial nutrients, phytochemicals, and other ingredients that will lead to new pharmaceuticals.

Pharmaceutical products are found in temperate rainforests as well. In the ancient forests of the Pacific Northwest, the Pacific yew *(T. brevifolia)* is the source of *Taxus baccata,* the active ingredient in the anticancer drug paclitaxel, which inhibits cell division. Once thought to be mere weeds or scrub growth, yew trees were usually burned with the undergrowth cut as loggers made their way to major stands of timber. But researchers discovered as early as the 1960s that the chemical in the bark could be an anti-tumor drug. In 1984, the National Cancer Institute began clinical trials that looked at paclitaxel's safety and efficacy in treating certain cancers. Other studies followed and by the end of 1992, the U.S. Food and Drug Administration (FDA) approved the use of paclitaxel for ovarian cancer.

Many environmentalists and some government officials feared that because of high demand for its bark, the Pacific yew would be quickly harvested out of existence, further endangering the old-growth forests that are the trees' primary habitat. But in 1992, the U.S. Congress passed legislation requiring long-term conservation and efficient management of the yew in national forests. New technologies were also developed to synthesize paclitaxel from every part of the tree—not only from bark, but also from twigs and needles. Beginning in 1992, the FDA approved the drug for treating breast cancers, and later for treating ovarian cancer and common forms of advanced lung cancer.

Indigenous People

Of the 250 million indigenous people worldwide, about 50 million make rainforests their home. Indigenous people in the rainforests make up more than 1,000 distinct groups. There are 700 indigenous communities in Papua New Guinea alone, and about 200 make Africa's Congo Basin their home. Other indigenous groups include the Penan of Sarawak, Malaysia; the Efe of Congo; the Lua of northern Thailand; the Hanunoo in the Philippines; the Kayapo and Yanomami of Brazil; the Sanema of Venezuela; the Lacandon Maya in Mexico; the Kuku-Yalanji of northeastern Queensland, Australia; the Pygmies of Central Africa, and the Desana of Colombia.

The worldviews of indigenous people are very different from those of the Europeans who have invaded their lands. As the Rainforest Foundation of Great Britain explains:

> For many indigenous peoples, land is held collectively, and according to indigenous law, humans can never be more than trustees to the land, and are perceived to have a collective responsibility to preserve it. There is often no concept of land ownership. Respect for Nature is embodied in the spiritual and social lives of indigenous peoples. On the other hand, the predominant Western view is that Nature should be studied, dissected and mastered, that land and resources can be owned by an individual or a company, and that "human progress" is measured by the ability to extract natural resources and accumulate material wealth as a result of the natural loss. . . .
>
> Traditionally, indigenous peoples practise shifting agriculture, hunting, and gathering, and fishing. They use an intimate knowledge of plants, soils, animals, climates, and seasons, to exploit nature in a way in which the environment can sustain itself as well as their needs. Their practises involve careful management, control of population, the use of small quantities but a wide diversity of plants and animals, small surpluses and mini-

mum wastage. Plants provide food, medicine, pesticides, poisons, building materials, and provide meat, clothes, string, implements, glues and oil (The Rainforest Foundation n.d.).

Over the centuries, millions of indigenous people in tropical rainforests have been killed by military troops and others who supported efforts to seize native lands for such activities as mining, oil drilling, and ranching. An estimated 10 million Indians lived in the Amazonian rainforest five centuries ago, but now there are fewer than 200,000. Since the 1900s in Brazil alone, European colonists have destroyed more than ninety indigenous communities. Indigenous people have also suffered illnesses and death due to pathogens and toxins introduced by Westerners and others outside their culture.

Works Cited

Ecotrust and Conservation International. 1992. *Coastal Temperate Rain Forests: Ecological Characteristics, Status and Distribution Worldwide.* Portland, OR: Ecotrust; Washington, DC: Conservation International.

National Park Service. August 18, 2000. "Rain Forest—Olympic National Park." http://www.nps.gov/olym/edurain.htm (accessed February 24, 2001).

The Rainforest Foundation. No date. "Indigenous People." http://www.rainforestfoundationuk.org/rainhome.html (accessed February 24, 2001).

World Rainforest Movement. 1992. *Rainforest Destruction: Causes, Effects and False Solutions.* Penang, Malaysia: World Rainforest Movement.

2

Problems, Controversies, and Solutions

Disappearing Forest Cover

Throughout the 1980s and 1990s, various organizations provided estimates on the amount of all types of forests—not just rainforests—that have been destroyed or degraded over the centuries. One estimate often used is that four-fifths of the forests that covered the earth 8,000 years ago have been destroyed or degraded by human activities—most of that since the 1970s. Another estimate says that forest cover—land area covered with forests—has been reduced by 20 percent or perhaps as much as 50 percent since preagricultural times.

Since 1980, forest area has increased slightly in industrialized countries but has declined by almost 10 percent in nonindustrialized countries. Tropical deforestation probably exceeds 80,600 square miles (130,000 square kilometers) a year. In one country alone, Mexico, half of the tropical and temperate forests have been lost since the 1960s, and Mexico's "last great reserves are under siege," according to a report in *Sierra* magazine. "The Lacandon jungle in Chiapas has been reduced from 5,000 square miles to less than 2,000, and loses another 125 square miles every year" (Ross 2000).

David Sandalow, U.S. Assistant Secretary of State for Oceans and International Environmental and Scientific Affairs in the Clinton administration, pointed out the global impact of this deforestation. In an address before the National Press Club, he stated that during the 1990s, "the world . . . lost an average of 38 million acres of forest per year. This is a land area roughly the size of Georgia. Tropical forests are vanishing at the rate of 250 acres per minute. To put that in context: a football field is roughly two acres. We are losing two football fields of tropical forest every second" (Sandalow 2000).

13

In 1997, the World Resources Institute estimated that about one-fifth of the world's frontier forests or large tracts of intact forest remained. But two years later, in 1999, the institute found that a "combination of satellite imagery, geographic information systems (GIS), mapping software, the Internet and on-the-ground observation" presented an even more devastating picture of the disappearing forests. According to Dirk Bryant, director of Global Forest Watch, a World Resources Institute project,

> It now looks like we have underestimated those threats in some places. The maps and reports reveal widespread logging in the forests of the Congo Basin, and extensive mining, energy and road construction projects in the forests of Canada. For the first time, we are mapping out logging across the Congo Basin, a region that contains the world's second largest contiguous tropical forest after the Amazon. Canada's intact forests are being opened up for large-scale exploitation, including those in the environmentally-sensitive far northern forests (World Resources Institute, 2000).

At the end of 2000, a report issued by the World Resources Institute entitled "Pilot Analysis of Global Ecosystems: Forest Ecosystems" showed that forest areas in developed countries increased slightly, while in developing countries, clearance for agriculture, development, and logging reduced forests by at least 86,800 square miles (140,000 square kilometers) annually. The report analyzed not only forest areas but also the ability of forests worldwide to provide a broad variety of goods such as timber, fuel, food, and medicines and such environmental services as water purification, carbon storage, and rich habitat for biodiversity. Demands for timber, fuel, mineral resources, and food production alter the distribution, density, and size of trees, and radically affect many other species that depend on forests, the report noted. The authors recommended that governments aggressively encourage timber production from plantations and forests that are intensively managed in selected areas, and discourage old growth harvesting (WRI 2000).

Government Policies Contributing to Deforestation

In both industrialized and nonindustrialized nations, government policies have frequently encouraged deforestation. Because of huge debts incurred during the 1980s, industrializing countries have supported development strategies designed to produce crops for export, earning funds to pay off loans and to buy manufactured goods. At the same time, industrialized nations have indirectly encouraged deforestation through their consumer demands for wood products, oil, and other natural resources from the rainforests.

Because of government encouragement, most of the fertile land areas in Amazonia and the Philippines have been turned into huge plantations, many owned by multinational companies, to grow such export crops as bananas, pineapples, sugar cane, rice, palm oil, coffee, and rubber. Major timber companies also develop tree plantations, which have little in common with forests. Thousands of trees of the same species are grown rapidly to produce the greatest amount of raw materials in the shortest time possible. Tree plantations require intensive preparation of the soil, fertilization, mechanical or chemical weeding, pesticide use, thinning, and mechanized harvesting. As a result, soil fertility and groundwater reserves are reduced, erosion increases, natural biodiversity is lost, and fire risks increase.

Many *peasants* (poor farmers who own no land) have little or no access to fertile soil, and so they use *swidden* agriculture practices that they have followed for centuries, cutting or burning forests and clearing the land to grow their own food. These farmers are frequently blamed for destruction of rainforest areas, but when they properly burn and farm small plots of land, they allow soils that have been recently farmed to regenerate and they also tend the forest habitats around them. Resettlement programs, however, have moved some peasants into already logged areas, where they try to farm the land only for the short-term gain. When the soil degrades within a few years and little grows, peasants move on to other land, repeating their "shifting cultivation" practices again and again just to survive and damaging more and more habitat.

Amazonia Rainforest Destruction

There are many opinions about how the destruction of the Amazon rainforest began. But in the opinion of Susanna Hecht, an expert on the Amazon, "[t]he causes of environmental degradation in the Amazon can be traced to a philosophy and strategy for regional development formulated by the Brazilian military," which gained great power and influence beginning about the 1930s and expanding rapidly after World War II. The Brazilian military pressed for the transfer of public lands to private landholders—primarily those with wealth and power—and for the development of roads through rainforest areas, displacing indigenous people and killing thousands through bloody conflicts (Hecht, 1990, chapter 6).

A military junta took over the Brazilian government in 1964 and began to implement plans to settle Amazonia, which made up half the nation's land area but was home for only 3.5 percent of Brazil's population of 70 million. By encouraging people in densely populated areas of Brazil to colonize Amazonia, the military government, with the support of churches and other organizations, hoped to provide land and "vibrant opportunities" for the Amazonian colonists. In the words of Hecht, church leaders "exhorted their charges to seek the promised land" and the Brazilian government "offered tangible inducements: transportation to the Amazon and a 240-acre plot for each settler, with sure title, guaranteed credit for the planting of rice, corn and beans; a six-month household subsidy to tide the family over the initial difficult months; and food subsidies as insurance against disasters. Colonists were also promised housing, schools, medicine, transportation and technical assistance" (Hecht, 1990, 124–125). But the farming efforts of many colonists failed because crops were diminished by poor soil or were destroyed by pests.

Beginning in the late 1960s, the Brazilian government implemented a plan to build roads and encourage commercial investment in the Amazon, offering tax incentives such as a moratorium on corporate income taxes for a dozen years, subsidized loans, and reduced import duties for manufacturing or agricultural equipment needed to develop Amazonian land. In spite of protests by Brazilian environmental groups and some Brazilian government officials and investments in conservation planning by international and national organizations, development projects have continued in Amozonia over the decades. In 2001, the Brazilian government announced its $40 billion Avança Brasil (Advance Brazil) program to accelerate industrialization, including new highway construction and hydroelectric projects. Research scientists writing in *Science* magazine predicted that the massive program will destroy or degrade 28 to 42 percent of the Amazon rainforest by 2020 (Laurance, et al. 2001 p. 438).

North American Policies

Temperate rainforests also have been greatly affected by govern-ment policies. North America's temperate rainforests once cov-ered over 60 million acres (25 million hectares) along the Pacific coast, less than half of which now remains. Since the mid-1800s, when people began to push West to exploit the natural resources of the area, lumber companies have harvested most of the timber in the rainforest areas. Logging increased substantially after World War II to meet the demands of a growing population for housing, businesses, and other construction purposes. Agricultural and urban development also contributed to forest losses. Since then, government policy has had a mixed impact, sometimes protecting the forests and sometimes leaving them vulnerable.

About 5 million acres (2 million hectares) of ancient forest, making up much of the coastal rainforest in the United States, are protected in parks and wilderness areas. Some of the forest areas are set aside to preserve the habitat of endangered species such as the spotted owl. Habitat preservation is required by the fed-eral Endangered Species Act of 1973, which has created contro-versy in many parts of the United States, especially in the Pacific Northwest where temperate rainforests are located.

The Endangered Species Act has become a target for groups aligned with the so-called wise use movement, which began in 1988 under the leadership of Ron Arnold, a journalist and former activist with the Sierra Club who now frequently condemns en-vironmental organizations, and Alan Gottlieb, president of the Center for the Defense of Free Enterprise. Arnold adopted the term "wise use" because of its association with Gifford Pinchot, the first chief of the U.S. Forest Service, who believed that forests and other natural resources should be used wisely to conserve them for future generations. Today's wise use movement, how-ever, is made up of many private property rights advocates and industries concerned that their land is being taken over by the government to preserve the environment at the expense of the local, state, or national economy. Wise-use groups include those involved with such industries as logging, mining, cattle ranch-ing, trucking, and wood products manufacturing and associa-tions concerned with recreational activities like hunting, fishing, and motorcycling in forests.

In 2000, Arnold appeared before the U.S. House of Representatives Committee on Resources, Subcommittee on Forests and Forest Health to charge that "the environmental movement is a three-cornered structure beginning with tax-exempt foundations which devise multi-million-dollar environmental programs to eliminate resource extraction industries and private property rights. The foundations direct their funds to the second leg of the triangle, environmental groups with insider access to the third leg, executive branch agencies. This powerful 'iron triangle' unfairly influences federal policy to devastate local economies and private property" (House 2000).

Gregg Easterbrook, a contributing editor to the *Atlantic Monthly* and to *U.S. News & World Report*, frequently argues in his books and articles that economic development, whether in the form of logging or another industry, is not necessarily evil. He points out that the Endangered Species Act has led to "a fierce backlash from landowners, who dispute the way the ESA sometimes forces them to set aside private property for use as animal habitat." In 1998, Easterbrook suggested a possible solution that would

> allow economic growth, and still protect land and species—perhaps more effectively than the current law does. Congress should simply require that, for every new acre of land converted into development, one acre of land elsewhere must be purchased and dedicated to conservation of habitat. The land purchases would be financed by a per-acre fee charged to all building on previously undeveloped land. That means all development, regardless of whether endangered species are directly affected. The fee would be assessed automatically in conjunction with building permits, obviating the need for complicated lawsuits and impact studies. In return, developers would be freed from most current ESA strictures. The utopian goal of protecting all species everywhere would be dropped from the act, though the government would retain emergency authority for a few rare cases (Easterbrook 1998a).

While the Endangered Species Act and forest preserves continue to draw controversy, U.S. laws do allow logging on some federal land, and a share of the funds from the sale of that timber is turned over to counties where national forests are located. The payments compensate the counties for property taxes that might have been paid by private owners. Other funds for maintaining and managing national forests come from such forest uses as mining, livestock grazing, and recreational activities, as well as from receipts for the sale of salvage timber and compensation for building roads for private timber companies. Most of the funds collected from timber sales are retained by the Forest Service. But because timber and timber resources are frequently sold at below market rates, timber sales do not cover all Forest Service expenses for those sales. In fact, from 1992 to 1997, taxpayers lost $2 billion on the Forest Service logging program, according to the U.S. General Accounting Office. In addition, most of the 440,000 miles (71,000 kilometers) of logging roads constructed in national forests were paid for not by the logging companies, but by the American taxpayers.

Timber industries in the Pacific Northwest have long argued that jobs depend on increased logging. However, according to a 1994 study by the Economic Policy Institute, job losses in the Pacific Northwest have been caused by such factors as automation within the industry, increased exports of raw logs, and unsustainable harvests during the 1980s. From 1980 to 1988, nearly 14,000 jobs were lost due to such factors as these:

1. Giant companies or financiers have taken over smaller firms and have sold off timber as fast as possible to pay off debts and provide funds to invest in other types of industries that show better profits than lumbering.
2. Many lumber mills have automated their operations, reducing the need for workers.
3. Rather than milling logs in U.S. mills, companies are shipping whole and split logs to Japan and other Asian countries and to Mexico—in effect exporting thousands of jobs along with timber.
4. The demand for timber from the Pacific Northwest has been decreasing since the early 1960s because lumber companies have been able to buy timber from other sources, some of which are woodlands elsewhere in the nation that have been planted specifically for the timber produced.

Arguments for and against logging in the Pacific Northwest have become increasingly strident in Alaska and British Columbia, Canada. Every year the U.S. Congress appropriates about $30 million to subsidize logging in Alaska's 17-million-acre Tongass

Rainforest, and logging has been vigorously defended by congressional members from Alaska. In 1998, Representative Don Young declared: "The economy of every community in Southeast Alaska depends directly or indirectly on the forest, including fishing, tourism, and yes, timber. The diverse uses the forest provides to local economies are like the legs on a three-legged stool—remove one and the whole thing falls. . . . In 1990, there were 4,225 Alaskans employed in the timber industry. Today [1998], over 65 percent of those jobs are gone" (Young 1998).

Young and other government officials as well as timber company representatives contend that logging should be increased in the Tongass so that the state can develop other industries and become more than "a federal park or forest." To that end, officials want to build 400 miles of more roads in the forest, adding to the 4,000 miles of logging roads that already exist. Yet a federal plan announced in late 1999 "would ban road building in areas of the national forests where there are currently no roads," *The New York Times* reported. "The plan would protect about 43 million acres that could be logged, roughly one-quarter of the 192 million acres in the national forest system. That would reduce planned timber harvests, now at about 3.6 billion board feet a year, by about 700 million board feet" (Verhovek 2000).

The federal proposal at first exempted the Tongass from the road ban because of a management plan, initiated in 1997, to balance logging with wilderness conservation. But that exemption was opposed by numerous environmental groups, and in late 2000 the federal government announced that protection would be expanded to include Alaska's Tongass National Forest.

Two large mills in southwest Alaska have closed because of reduced logging, but tourism has increased. *The New York Times* reports that "many of those who make their money by guiding tourists and sport fishers through the spectacular inlets and past the emerald forests favor more protections, and even an end to any clear-cutting in the Tongass" (Verhovek 2000).

In British Columbia, a similar controversy prevailed during the 1990s over temperate rainforests on the central and north coasts, known as the Great Bear Forest. A coalition of environmental groups, including the Coastal Rainforest Coalition, Greenpeace, Rainforest Action Network, and the Sierra Club of British Columbia, has used boycotts, blockades of logging roads (some involving violent confrontations), and letter writing campaigns to prevent timber cutting in biologically rich and untouched parts of the Great Bear Forest. Within this area of approximately 7 million hectares are territories of indigenous groups whose rights and title to the land have not been resolved.

In the summer of 2000, the environmental coalition announced that it and four regional coastal forest companies—Canadian Forest Products, Fletcher Challenge Canada, Western Forest Products, and Weyerhaeuser—would work together to develop a conservation-based ecosystem approach to managing the temperate rainforests on the central and north coasts of British Columbia. The companies and the environmentalists agreed to establish an independent process for identifying social and economic impacts associated with the suspension of logging and to work out alternatives to maintain employment stability while a proposal on ecosystem planning is developed.

Logging in the Tropics

In nonindustrialized nations of the tropics, wood is the primary energy source for millions of poor and landless rural people. Many peasants cut down trees to attain the fuelwood they need for cooking and heating. Fuelwood harvesting depletes 5 million acres (2 million hectares) of tropical rainforest every year. The timber industry has long claimed that rainforest losses are due to fuelwood harvests and land clearing by the increasing number of swidden agriculturists, but commercial logging has had a much more devastating impact and is a primary agent responsible for tropical deforestation. Indeed, the two types of forest destruction often go hand in hand; along with the direct impact of cutting down trees, commercial logging has an indirect impact because it involves the building of roads. Landless farmers then use the roads to gain access to rainforest areas that they clear by slashing and burning. The cleared land is used to grow enough food to keep families alive.

Logging is also the source of timber exports, which provide income for many nations in tropical regions. Tropical woods have been shipped primarily from Africa and southeast Asia to Japan, the largest consumer of tropical timber. Japan uses much of its imported tropical timber to make such products as nonreusable wooden forms for concrete construction, plywood, furniture, chopsticks, and toothpicks. The country also uses wood chips to make cardboard packaging for electronic equipment. Originally those chips came from the United States, but since large quantities of U.S. waste chips are no longer available, Japan set up its own operations in the island nation of Papua New Guinea (PNG), clearing hundreds of acres of rainforest on its north coast.

It has been a long-standing policy of the PNG government to accommodate transnational timber companies. But in 1999 the government imposed a logging moratorium, primarily because the World Bank, which provides loans for such economic development projects as logging and dam building, demanded reforms and a thorough review of logging operations before releasing funds. But in early 2001, the World Bank appeared ready to turn over funds for economic development in Papua New Guinea even though a review of forestry operations had not been completed. A coalition of nongovernmental organizations contends that "the World Bank's rhetoric on eradicating poverty, supporting good governance and protecting the environment are not being translated into effective measures. . . . Indeed in Papua New Guinea, it seems that the World Bank is pursuing actions that are going to compound poverty, endorse corruption and further environmental destruction" (forest.org 2001).

In Malaysia, efforts have been under way for years to stem extensive damage to and destruction of rainforests, particularly in Sarawak. Since 1989, the Penan, an indigenous group of hunters and gatherers sustained by the rainforest plants and animals, have tried to prevent further logging by barricading roads into the tropical forests. But government officials have arrested many of the Penan people, and some leaders have had to flee the country. Most of the Penan, altogether numbering about 9,000, live in the interior forestlands, but many have been forced to live in villages and cities because their homelands have been devastated.

In mid-2000 more than 100 Penan men and women staged a nonviolent protest on a logging road leading to their communal lands in the Baram region of Sarawak. The Penan erected a wooden barricade to prevent the logging companies from taking timbers from their lands. According to numerous news reports, the Penan resorted to this action because their rights of access to their natural forest resources were continually ignored by logging companies. The Penan's numerous complaints to the authorities and the logging companies regarding their claims to the forest resources and the problems caused by logging have fallen on deaf ears. They believed they had no other alternative except to stage a protest and vowed to continue blocking the road until their problems are resolved.

Vast amounts of rainforest in the tropical regions of U.S. territories also have disappeared because of logging and highway construction. For example, rainforests on the Virgin Islands were destroyed years ago for sugar and cotton plantations and cattle ranches. The forests have grown back partially but now are threatened by home construction and resort development. Tourism and resort development also threaten rainforests on the islands of American Samoa, although the U.S. Congress designated two of the islands to be part of a new national park in order to protect forest and reef areas.

Other Causes of Tropical Deforestation

In the tropics, other human activities besides excessive logging contribute to deforestation. These include the development of rubber, oil palm, sugar, banana, coconut, coffee, and tea plantations; programs to relocate growing urban populations to fringe forest areas; road construction; cattle ranching; urban development; hydroelectric dam construction; oil extraction; mining; and shifting cultivation. Some of these practices are intertwined; for example, road building is needed for dam construction, mining, and oil extraction, and all of these cause deforestation. Several of the various activities that contribute to rainforest losses are described in the sections that follow.

Cattle Ranching

In South and Central America, cattle ranching is a primary cause of deforestation. Ranchers clear rainforest land to plant pasture grass for their cattle, but as is the case with some farming methods, the soil is soon depleted, and so more rainforest land is cleared. Much of the beef produced in Central America, particularly during the 1970s and 1980s, was exported to North America for consumption as hamburger in fast-food restaurants, in frozen food meals, and for dog food. During the 1990s, some U.S. fast-food chains such as Burger King and McDonald's declared that to preserve rainforests, the companies would not use Central American beef or beef raised on lands that were once rainforests.

Wealthy landowners and international corporations and banks are the main beneficiaries of cattle raising, taking over the productive land and leaving marginal land for poor rural families.

The environment pays a heavy price; according to a report issued in 1997 by the Food and Agriculture Organization of the United Nations (UN),

> The transformation from tropical forest to crop and pasture land brings about substantial losses of soil fertility and soil erosion. Furthermore, in many tropical rainforest areas, pastures can only be sustained for a short period of up to ten years. Soil nutrients are rapidly depleted after clearing and grasses are soon replaced by less useful vegetation. Natural regeneration of forests is quite difficult, especially in large degraded areas. More than 50 percent of the pasture areas in Amazonia have now been abandoned in a degraded state (de Haan, Steinfeld, and Blackburn 1997).

Oil Extraction

Drilling for oil has endangered diverse rainforests, including those in the Americas and Africa. Dozens of indigenous groups, among them the U'wa of Colombia, the Karen of Burma, and the Nahua of Peru, are also threatened by the worldwide expansion of the oil industry. Numerous international environmental organizations and some indigenous groups have tried to stop further oil exploration.

Although a large portion of the Ecuadorian rainforest, including Yasuní National Park, has been set aside as a reserve, oil companies have ravaged about one-tenth of the forest since the 1970s, drilling hundreds of wells and constructing a network of roads and pipelines. A 280-mile pipeline carries heavy grade crude oil to ports, where it is pumped into tankers and transported to refineries around the world. At least half of the exports go to the United States.

For a time, the Huaorani people stopped the Ishpingo, Tambacocha, Tiputini, Imuya oil drilling project, called ITTI, within the Yasuní National Park, and part of the territory is now protected from oil exploration. In addition, the Cofan people shut down the Dureno well in Amazonia, preventing the extraction of some 1.27 million barrels of oil.

In Colombia, heated and often violent controversies over oil exploration and drilling have been occurring since the 1970s. Guerrilla groups see oil industry projects as government installations and therefore targets for attacks. As a result, the government has militarized oil production. The constant fighting has taken a toll on many indigenous peoples of Colombia, including the Yarique, the Kofan, and the Secoya.

The U'wa, an indigenous community of 5,000 in Colombia, have long seen the devastating impacts of oil projects near their homeland. Since 1986 an estimated 1.7 million barrels of oil have spilled from the nearby Caño Limon pipeline belonging to U.S.-based Occidental Petroleum, polluting soil and waterways. Oil projects also have led to deforestation as land is cleared to construct roads and lay pipelines. The roads then encourage poor farmers to come into the forest and clear even more land for settlements.

In 1992, the Colombian government gave Occidental permission to drill for oil on land that the U'wa have survived on for thousands of years. The Colombian Constitution of 1991 provides indigenous peoples with the right to be consulted about oil extraction and mining in nearby territories. The land under dispute is supposedly outside U'wa territory, but it is part of their traditional migratory land, which once covered more than 3 million acres. The U'wa claimed the right to veto the project.

In 1997, the Colombian Petroleum Association contended that indigenous peoples had erroneously interpreted their consultation privileges to mean they have veto power. In the same year, a court decision supported the oil company, giving Occidental permission to drill in about 400,000 acres near a pipeline it had completed in 1985 (Kraul 1997).

However, soon thereafter an international coalition of environmentalists, including several U.S. groups, began campaigns to support the U'wa cause, and the U'wa themselves have continually protested with roadblocks and legal action, declaring they would commit mass suicide before they would allow destruction of their land and culture.

In the spring of 2000, a Colombian judge ordered Occidental to stop all work on its proposed well site, but a high court revoked the injunction just six weeks later. The outcome of this dispute is still uncertain, as protests against Occidental continue and the Colombian government holds firm in its support of drilling and the export of oil.

Dam Building

Hundreds of thousands of square miles of rainforest have been lost due to dam building and flooding for reservoirs that provide hydroelectric power. Governments in countries such as India, Indonesia, and Brazil have encouraged dam building in order to support new industries, including mining and foundries, and to provide irrigation. For years, the World Bank has provided loans for dam building, contending that the development would improve economic conditions. But the dams and industrialization have brought few benefits to the rural poor or indigenous groups. Dam construction requires the displacement of people and communities, which in turn destroys cultures and condemns many to life in urban slums.

In Honduras, plans to build hydroelectric dams on the Patuca River have created great controversy. The river region includes the tropical forest homes of four indigenous groups—the Miskitos, Tawahkas, Garífuna, and Pech—who depend on the fish in the river for part of their food supply. One group, the Tawahkas, would have to be relocated if the dam were built, as the dam would flood their forest homeland. Cutting trees and building roads for access to dams could also encourage subsistence farmers to settle in the area, slashing and burning forests to plant crops. Roads also open the way for commercial interests—logging and cattle ranching among them—to take over the land.

On the other side of the argument are those who say that the dams would help create needed jobs and would encourage tourism, helping to improve the economy of the country. In addition, proponents contend that dams are needed to meet the ever-growing demand for electricity—only 20 percent of Honduras has access to electricity. The increased energy demand is related to the *maquiladora* system, in which workers in Honduras assemble goods from parts made in the United States, Japan, or another industrialized nation. The assembled goods are then shipped back to a *maquila* (sister) packaging plant in the country of origin. Honduras also would like to take advantage of opportunities to sell electricity to other Central American countries.

In 1998, the World Bank and the World Conservation Union helped launch an independent commission called the World Commission on Dams to review the effects of 1,000 large dams worldwide. (Of the estimated 800,000 dams around the world, about 45,000 are considered large dams.) The twelve-member commission included industry representatives, dam owners, government officials, and environmentalists.

In its report, released in November 2000, the commission concluded that even though dams create such benefits as hydropower and inexpensive irrigation systems for agriculture in developing nations, the overall impacts on people and ecosystems have been "unacceptable and often unnecessary" in too many cases. The commission's 400-page report estimates that dam projects have displaced between 40 million and 80 million people and have been responsible for the loss of floodplains, forests, and biodiversity. As an overview of the report, available in its entirety on the Internet and in printed form, points out:

> The controversy over dams has appropriately been raised to the international stage. A dissipation of that controversy, however, should allow decisions about fundamental water and energy development choices to be made at the most appropriate level—one where the voices of powerful international players and interests do not drown out the many voices of those with a direct stake in the decisions. For this to work, all the actors have to make a commitment to step out of their familiar frames of reference (World Commission on Dams 2000).

The commission recommended that its report be used to encourage discussions, debates, and reassessments of established procedures. Perhaps then changes could be made to reflect the needs of the people most affected by new dam construction.

The Impacts of Deforestation

Indigenous People in Jeopardy

Approximately 250 million indigenous people live in 70 countries and control 12 to 19 percent of the earth's land. Each indigenous community that lives in a tropical or temperate rainforest area has its own way of life, but all rely on the forest for some if not all aspects of their survival. Indigenous rainforest communities depend on the planet's biological diversity, and in turn, the planet's health depends on indigenous people because they have strong connections to the land and maintain traditions that help protect ecosystems.

Consider the Efe and Lese peoples in the northeastern part of the Democratic Republic of the Congo. They live in relative isolation in the Ituri tropical rainforest. Cultural Survival, an organization that helps isolated people survive and adapt after contacts with the outside world, describes the relationship between these groups and their surroundings:

> The Lese live in small permanent villages and practice slash and burn agriculture. The Efe are nomadic foragers who move around the forest to hunt animals and gather plants, honey and insects. The Efe not only eat these forest foods, they also trade them with the Lese for cultivated foods and a few essentials like cloth, salt and iron.
>
> The Efe and Lese are barely assimilated into their national culture and they are rarely involved in important decisions affecting their lives (Cultural Survival n.d.).

Because of deforestation, many indigenous cultures and the very lives of indigenous people have been threatened over the past several decades. In Brazil, the livelihood and lives of indigenous groups were in jeopardy when the national government planned to flood up to 97,000 square miles (250,000 square kilometers) of land for dam projects to be financed by loans from the World Bank. Out of 168 planned dams, 68 were on territories claimed by indigenous groups. But environmentalists in Brazil and other countries encouraged leaders of the Kayapo to launch a campaign during the 1980s to stop the dam building projects.

Kayapo leaders traveled to the World Bank headquarters, where they asked officials not to fund the dam building. They also met with federal legislators in the United States and Europe. In addition, the Kayapo held an international meeting in 1988 at one of the dam sites, which resulted in worldwide publicity and pressure on Brazil to withdraw its loan proposal, halting the dam project.

For indigenous people in many countries, contact with outsiders—usually whites of northern European extraction—continues to have fatal consequences. Many people die from contagious diseases for which they have no immunity, ranging from measles to influenza to sexually transmitted diseases. More than 30 percent of the Yanomami people in Venezuela, for example, have died from whooping cough and measles since their first contact with outsiders. Air, soil, and water pollutants from industries established by Asian, European, and American companies in tropical forest areas also have posed serious health hazards for indigenous groups in addition to harming the rainforests.

For example, not only does oil exploration destroy vast amounts of rainforest, but it also contaminates soil and drinking water supplies. Extremely toxic wastes from oil production are dumped into open pits, creeks, and streams, threatening the health and lives of indigenous people. One investigative reporter visiting Lago Agrio, a so-called boomtown near Amazon oil country, describes the town as "a full-blown slum in the midst of what was only twenty years ago virgin jungle. There are open sewers, rooting pigs, piles of garbage on the unpaved rutted streets . . . dilapidated housing, sooty tire-recapping stands, used auto-part stores . . . hookers displaying their thighs in open doorways, two-dollar-a-day oil workers knocking back their second six-pack by noon, and a prevailing mood of restlessness, tension, and frustration" (Cooper, 1992, 44).

People in so-called advanced nations can learn much from the world's Native peoples about how to use the land and harvest products without destroying the environment. Consider the Kayapo of Brazil's rainforests who plant their gardens in complementary groupings, a practice now popular in industrialized nations. Certain plants thrive when growing beside another type of plant; other plants provide protection to neighboring plants. For example, some plants release substances that discourage insects, protecting not only themselves but other plants nearby. The Kayapo also fertilize their garden plots with rotting plants and soil from ant and termite nests—all rich in nutrients. In addition, the Kayapo gather and make use of hundreds of plants for medicinal purposes.

Wherever the culture and lives of indigenous people are extinguished, thousands of years of accumulated knowledge of natural ecosystems are lost. This in turn affects not only the health of the environment, but that of all people in industrialized societies who could have benefited from this vast store of knowledge.

Effects on Natural Resources

Deforestation is responsible for the destruction of forest products such as fruits, nuts, and various fibers, and the loss of valuable medicinal plants. Scientists from the New York Botanical Garden, the Smithsonian Institution, and elsewhere have compiled a "Red List" of endangered plant species, which estimates that one in eight plant species in the world, or nearly 34,000 species, are threatened with extinction.

One of those plants is the rosy periwinkle, which grows in

the ancient forests of the island of Madagascar, off the southeast coast of Africa. The rosy periwinkle is known throughout the medical and scientific communities worldwide because the plant's chemicals are used effectively to fight childhood leukemia and Hodgkin's disease. These chemicals are also used in medications that lower blood sugar levels, slow down bleeding, and tranquilize. According to the National Wildlife Federation,

> stable wild populations of rosy periwinkles still grow in Madagascar, [but] mass deforestation due to poor agricultural practices is rapidly depleting the island's native dry tropical forest habitat and threatening its unique plant and animal species. Slash-and-burn farming has been standard practice for many years as a means of converting forest areas to crop fields. Some scientists estimate that at the current rate of destruction, almost all of Madagascar's forests could be lost within 25 years (NWF 2000).

Along with the diverse plant life that is being threatened, many mammal, bird, insect, and other species are also becoming extinct. Species loss has been estimated to occur at a rate of about 50 to 150 species per day, a rate that has not occurred since the dinosaurs became extinct 65 million years ago. The loss of individual species, from bacteria to mammals, has often been likened to the continued loss of the rivets that hold an airplane together. Although the short-term effects may not be noticeable, the long-term effects could be disastrous.

One scientist working to explain this high rate of species loss is Oxford ecologist Norman Myers, who originated the idea of "hot spots"—biological storehouses that face multiple threats. As an example, Myers points to the forests of northeastern Ecuador. "As recently as 1960," Myers writes, "they harbored at least 17,000 endemic species—that is, species found nowhere else—all crammed into an area no larger than Massachusetts. Within little over a decade, more than nine-tenths of these forests were cleared to make way for banana and other plantations" (Myers 1999).

Peter Raven, president of the International Botanical Congress and director of the Missouri Botanical Garden in St. Louis, has also compiled data on the extinction rates of plant and animal life around the world. Like the Red List, his data predicted that without drastic action, two-thirds of the world's 300,000 plant species will be lost during the next century as their habitats are destroyed. Raven's plan to slow the extinction rates of plants around the world includes a suggestion that the United Nations sponsor a major conference on the topic and that a new coordinating body, perhaps within the United Nations, monitor the status of plants throughout the world, to detect those most in danger, and take steps to conserve them.

When a species becomes extinct, its genetic resources disappear too, which "diminishes humanity," in the words of world-renowned Harvard University entomologist Edward O. Wilson. According to Wilson:

> Every microorganism, animal and plant contains on the order of from one million to 10 billion bits of information in its genetic code, hammered into existence by an astronomical number of mutations and episodes of natural selection over the course of thousands or millions of years of evolution. Biologists may eventually come to read the entire genetic codes of some individual strains of a few of the vanishing species, but I doubt that they can hope to measure, let alone replace, the natural species and the great array of genetic strains composing them. . . . Species diversity—the world's available gene pool—is one of our planet's most important and irreplaceable resources. No artificially selected genetic strain has, to my knowledge, ever outcompeted wild variants of the same species in the natural environment (Wilson, 1989, 114)

The loss of species and their genetic codes can have an adverse impact on the world's agriculture. Many food crops grown in the United States and other industrialized nations have been improved through crossbreeding with tropical varieties of plants, making the resultant crops more resistant to disease.

Crop production is also enhanced with the use of tropical insects in biological pest control programs. Some insects, called *beneficial insects*, act as predators or parasites, destroying other insects that are harmful to crops. Because many harmful insects have become resistant to pesticides manufactured from artificial chemicals, in some cases the use of beneficial insects has proven to be the most effective way to reduce or eliminate pests. Natural pest controls also help cut down on the use of chemical pesticides. This is important because many of these pesticides pose health hazards to the farm workers who apply them and to people who use pesticide-contaminated water supplies or who eat foods covered with pesticide residues. When forests are destroyed, many of the beneficial insects as well as plants disappear also.

Logging vast areas denudes the land, causing it to erode. Erosion increases the amount of sediment in rivers, in turn destroying habitats for fish and other aquatic life. Silt from the runoff of topsoil slows water flow and reduces the efficiency of hydroelectric dams. Sediment also adversely affects water quality. Finally, when land is stripped of trees, desertification can sometimes result—that is, the cleared land may become a desert.

Changes in rainfall could be another serious consequence of deforestation. In Panama, for example, the rainfall pattern over undisturbed forest areas has remained fairly constant. But in areas where forests have been cut, the rainfall has decreased steadily since the 1950s. Because of the forest's role in evapotranspiration, tropical deforestation could also result in decreased rainfall and droughts over much of the food-growing areas of the United States and the former Soviet Union.

These dire predictions notwithstanding, some researchers contend that the threat of imminent natural resource depletion is exaggerated. For example, Oxford scholar Wilferd Beckerman, former member of the Royal Commission on Environmental Pollution, argued in his 1996 book, *Through Green-Colored Glasses: Environmentalism Reconsidered*, that the only way developing nations will be able to deal with ecological problems is to industrialize and grow their economies. Only in this way will they be able to develop the technology and wealth needed to protect the environment. In short, Beckerman's position is exactly opposite that of rainforest advocates and other environmental activists.

Forest Fires and Global Warming

Record Forest Fires

In late 1997, tropical forests all over the world were burning. More fires were recorded in that year than at any time in history—millions of hectares of forest were destroyed in Indonesia and Brazil, along with vast areas of Papua New Guinea, Colombia, Peru, Tanzania, Kenya, Rwanda, the Congo, and other parts of Africa. Other large-scale fires burned in several Mediterranean countries and in Australia, Russia, and China.

According to a report from the World Wildlife Fund for Nature (known as the World Wildlife Fund in the United States), most of the fires were deliberately set, "often to clear land for planting, to cover up illegal logging and sometimes to open up land for development." The fires in 1997 "were worse because of the century's most severe El Niño weather event, which has caused prolonged droughts over much of the planet. El Niño events are growing more frequent and severe, probably as a result of pollution-induced climate change. The forest fires are turning previously moist forests into drier habitats that burn more easily as global warming begins to bite. Carbon dioxide and other gases released from fires add to the greenhouse effect." (WWF Press release, 1997).

Early in 1999, the fires started again, and parts of the Brazilian Amazon burned, as did areas of Manú National Park in Peru and regions of Bolivia and Paraguay. Fires also destroyed thousands of acres of forests in the United States and Russia. The World Wildlife Fund and the World Conservation Union, formerly known as the International Union for Conservation of Nature and Natural Resources, issued a follow-up report on forest fires, noting that although action has been taken to prevent forest fires, it has been "too slow and, in many cases, misdirected." The report issued a warning:

New evidence from the Amazon has concluded that fire causes a positive feedback cycle in which the more forests burn the more susceptible to future burning they become.This raises the possibility of large wildfire episodes happening on such a frequent scale that the forest ecosystem will not endure. The scientists believe the whole Amazon itself is threatened, with the rainforest being replaced by fire-prone vegetation. This has global consequences for biodiversity and climate change. (Rowell and Moore, 2000, 5).

Climate Change

Climate change—or global warming—has long been linked to deforestation. Many scientists theorize that not only large-scale burning of forests but also extensive logging in rainforests releases large amounts of CO_2, which interacts with other chemicals in the atmosphere, creating a buildup of so-called greenhouse gases (GHGs). In 1996, global emissions of CO_2

reached a new high of nearly 26,338 million tons (23,900 million metric tons), nearly four times the 1950 total, according to the UN Environment Programme. This GHG buildup could increase the overall temperature of the planet.

Global warming could alter rainfall, wind, and heat patterns around the world. Already warming has affected polar ice. Research scientists reported in 2000 and 2001 that glaciers, or sea ice, in the Arctic are shrinking at a more rapid pace than seen in centuries. Ice atop Mount Kilimanjaro in Tanzania and peaks in the United States and the Swiss Alps are also losing volume, indicating increased global warming. If the polar ice caps continue to melt at an increased rate, sea levels could rise, which would flood many coastal areas, including some along U.S. coasts. Coastal flooding would endanger many people since some of these areas are heavily populated. The agricultural belts of the world could also be altered, severely disrupting the economies of food-exporting nations like the United States. Under the predicted rate of temperature change, many species would be unable to adapt. They might attempt to migrate to more suitable environments, but migration could be blocked by urban development and other human activities that destroy habitats.

International efforts to offset global warming have been mixed. Although 100 countries had signed the 1997 Kyoto Protocol by 2000 agreeing to reduce their CO_2 emissions relative to 1990 levels, no industrialized nations had signed the treaty. To enable countries to continue supporting their industries while fulfilling their Kyoto commitments, various economic schemes have been put forth. One suggestion is that extensive tree plantations might create "carbon sinks," which would remove CO_2 from the environment. But such a scheme is not as simple as it appears. As a 2000 World Rainforest Movement report stated:

> The reasoning seems quite straightforward: while trees are growing, they take carbon dioxide from the atmosphere and fix the carbon in their wood. They thus act as "carbon sinks" and therefore help to counter climate change by removing carbon dioxide from the atmosphere. . . . [But] tree plantations are not aimed at supplementing measures adopted to reduce the use of fossil fuels. On the contrary, their aim is to allow industrialized countries to meet their reduction commitments without actually reducing them to the extent agreed upon. If, for instance, a country has made a commitment to reduce fos-

sil fuel emissions from 100 to 90 units, then instead of reducing by 10 it could reduce by only 5 and plant trees to absorb the remaining 5 (WRM, 2000 [Bulletin #37], 3–4).

The science of global warming also complicates things. For example, a renowned expert on global warming, James E. Hansen, who has contended for almost two decades that nations should cut CO_2 emissions to slow global warming, concluded in 2000 that other GHGs should be cut first. Hansen and his colleagues at the NASA Goddard Institute for Space Studies published a study in the *Proceedings of the National Academy of Sciences*. In their analysis, the scientists "suggest equal emphasis on an alternative, more optimistic, scenario that emphasizes reduction of non-CO_2 GHGs and black carbon during the next 50 years. This scenario derives from our interpretation that observed global warming has been caused mainly by non-CO_2 GHGs. Although this interpretation does not alter the desirability of slowing CO_2 emissions, it does suggest that it is more practical to slow global warming than is sometimes assumed" (Hansen et al. 2000).

Some scientists expressed concern that Hansen's new analysis would hinder efforts to reduce GHGs and that industries using fossil fuels would use the study to argue that the science of climate change is flawed. A number of climatologists and atmospheric scientists, such as William M. Gray at Colorado State University and Richard S. Lindzen at the Massachusetts Institute of Technology, have challenged the concept of global warming for years, contending that the rise in surface temperature—about 1 degree over the twentieth century—is natural and not significantly affected by human activities. These scientists contend that future warming could be beneficial for agriculture because growing seasons would be extended.

Journalists also express a variety of opinions on global warming. Gregg Easterbrook criticizes "orthodox environmentalists" who argue that Western nations should "immediately cut their emissions by 50 percent or more." He warns that such "a crash program against greenhouse emissions would almost certainly cause a recession, since it would require car-use bans, energy rationing, and similar strictures." In his view, an immediate priority should be developing strategies to cope with a warmer global climate: "If global warming theory really is right, climate change will arrive before even the most ambitious reforms could counter the buildup of greenhouse gases" (Easterbrook 1998b).

Both sides in the debate over climate change generally agree that Earth's surface temperatures and concentrations of GHGs are rising. According to a *New York Times* report, CO_2 concentration at the beginning of the twenty-first century "is nearly 30 percent higher than before the Industrial Revolution, and the highest in the last 420,000 years. Mainstream scientists, citing recent studies, suggest that the relatively rapid warming of the last 25 years cannot be explained without the greenhouse effect" (Stevens 2000).

The global warming debate will no doubt continue in the years ahead and scientists will continue to search for evidence to pinpoint the causes. At the same time, countless individuals, groups, organizations, and government agencies are convinced that preserving rainforests will also help slow global warming.

Efforts to Preserve Rainforests

During the 1990s, efforts to alert the public and to encourage action to preserve rainforests increased dramatically. This was in part due to rapidly expanding access to the Internet, which improved the ability of individuals, environmental and civil rights groups, and government agencies to spread their messages around the world at a rapid pace. In some countries, indigenous groups have also joined together to demand the reform of national policies that have destructive effects on rainforests. Some of these indigenous organizations have gained title to forest lands and have been managing them, as they have done for many thousands of years, so that the habitats will be preserved.

International Efforts

Since the early 1980s, deforestation and other forestry issues have gained attention on the international level. The UN Development Programme and the UN Food and Agriculture Organization made an assessment of tropical forests and along with the World Bank and the World Resources Institute developed a Tropical Forestry Action Plan that was implemented in 1985. It received strong support from at least eighty national governments representing 90 percent of the tropical forest area. But many nongovernmental organizations (NGOs) have criticized the plan, arguing that it does not take into account the effect of deforestation on climate change and biodiversity, nor does it acknowledge the need for involvement of local communities.

In 1990, the UN General Assembly established the framework for the UN Conference on Environment and Development, and for the next two years, organizers planned the agenda for a global environmental meeting. When the conference, dubbed the Earth Summit, convened in Rio de Janeiro in 1992, representatives of more than 150 national governments attended. Some participants in the Earth Summit hoped to reach global agreement on numerous environmental issues, among them how to counteract deforestation worldwide. During the summit, participants signed documents that included treaties on climate change and biodiversity, a statement of principles on general environmental protection (the "Rio Declaration"), and a Statement of Forest Principles, the first global agreement on the importance of forests and policies for conserving them. The Statement of Forest Principles is a nonbinding declaration, but it reflects global consensus on some forest issues and provides guidelines for conserving and managing forests in a sustainable way (both of these documents can be found in Chapter 5).

Other United Nations activities have included the establishment in 1995 of the Intergovernmental Panel on Forests to identify underlying causes of deforestation and forest degradation. In 1997 the panel issued a report that included 135 proposals for action that governments agreed to implement. Among them were proposals to complete in-depth studies of deforestation causes.

Every two years, the UN Food and Agriculture Organization publishes the *State of the World's Forests* to make current information available for policy makers, foresters, scientists, and others around the world. The publication includes data on changes in forest cover, trends in forest management, forest products, and international initiatives on forest preservation.

Government Actions

Various national governments have also taken action to stop forest destruction. The government of Bolivia, for example, joined with The Nature Conservancy and the American Electric Power Company in 1997 to sponsor a project to expand the Noel Kempff National Park and thus help reduce CO_2 emissions by an amount equal to the lifetime emissions of 500,000 cars. In another cooperative effort, the government of Suriname, the Inter-American Development Bank, the World Resources Institute, Conservation

International, and others worked together to protect 4 million acres (1.6 million hectares) of primary tropical rainforest in Suriname.

In 1998, the U.S. Congress passed the Tropical Forest Conservation Act, which authorizes a reduction of official debt owed the U.S. government by countries with significant tropical forests in return for forest conservation measures. At the end of 1999, President Clinton directed the Forest Service to devise plans to permanently protect an additional 40 million acres (16 million hectares) of national forest lands by banning road building.

But there is widespread concern among many environmental groups that the ban may be overturned by the administration of President George W. Bush who expressed opposition to his predecessor's directive.

Whatever the programs, "[t]here is no one-size-fits-all solution to protecting the world's forests," cautions David B. Sandalow of the U.S. Department of State. He warns against using scarce resources to negotiate an international convention on forests, which environmental organizations have been urging governments to adopt since the Rio Summit of 1992. "If we search for a single program, a single framework, or a single institution to comprehensively address this problem, we'll be disappointed" (Sandalow 2000).

Instead Sandalow suggests a dozen initiatives that would make a difference. For one, he urges the use of satellite imagery and new communications technologies to better assess, monitor, and manage the world's forests. He calls for more training of forest managers, foresters, biologists, and park guards:

> Managing forests wisely requires expertise and training, which are too often lacking in forest-rich countries. We must invest in people and communities, and we must train forest managers at all levels, including foresters, biologists and park guards, who can make a difference in the long-term. We also must educate people about the benefits forests provide and importance of conserving them. US government agencies from Agriculture to Interior to AID support training and education programs in countries around the world as do many environmental groups and businesses. But we need to see what more we can do, both bilaterally and internationally through organizations like FAO, ITTO (International Tropical Timber Organization) and the World Bank (Sandalow 2000).

Grassroots Efforts

Grassroots efforts to prevent rainforest destruction range from political campaigns and fund-raisers to front-line blockades to stop logging and road building in forests. In industrialized countries, consumer action to save rainforests also has been effective. In the United States, boycotts and campaigns have helped bring about pledges from home improvement companies to stop selling products, like construction lumber and plywood, that come from endangered forests. One of the largest suppliers of building materials, 84 Lumber Company, intends to phase out such forest products by the end of 2003. The company also plans to work closely with its suppliers to implement independent, third-party certification systems to inform and ensure consumers that their wood products come from well-managed forests.

Ecotourism has become a promising conservation effort in several South and Central American countries, including Belize and Costa Rica. In Brazil, a grassroots group working with the Institute for Social and Environmental Studies raised $250,000 from Anheuser-Busch and the U.S. Agency for International Development to preserve Una Eco Park in the Atlantic Forest. The park provides habitat for 450 different species of trees, the highest diversity in the world. Support for the park comes from ecotourism. Tourists from beach hotels pay ten dollars to walk across rope bridges 350 feet long, strung high in the trees. Some 6,000 visitors are expected in the park annually (Rogers 2000).

Some groups and individuals support debt-for-nature swaps, a program in which environmental organizations help pay off a portion of a nation's debt in exchange for rainforest land, which is set aside as a reserve. Other groups promote the sale of products from rainforests, which helps prevent the destruction of trees. For example, the environmental group Conservation International has initiated a variety of community development projects, including one in Ecuador to harvest and market *tagua*, an ivorylike palm nut that is used to make buttons, jewelry, watches, carvings, and other products. The *tagua* nuts are sold directly to factories, providing income for local workers and helping to create higher demand and prices for the nuts.

With the help of The Nature Conservancy, a group of Ecuadorian citizens was able to set up a nonprofit conservation organization and in 1988 purchase acreage for the Maquipucuna Nature Reserve, one of the last remnants of cloud forest—a tropical forest at high elevations often shrouded in clouds—in northwestern Ecuador. Maquipucuna lies in a region where rapid deforestation has been caused by timber, agriculture, pasture, and charcoal production. The original purchase of land consisted of 7,410 acres (3,000 hectares) of primary forest. With subsequent purchases of adjacent abandoned farms, the reserve now consists of more than 4,000 hectares and includes the Thomas H. Davis Ecotourism Center. The center generates funds to manage the reserve, provides work and income for local residents, and educates local visitors and international tourists about cloud forest ecology and the importance of conservation.

Visions for the Future

In spite of the many efforts to preserve the world's rainforests, there are numerous predictions that rainforests will disappear at a rapid rate in the years ahead. Yet some researchers and policy makers are cautiously optimistic that preservation and conservation programs will win out.

Education on the importance of rainforests is a top priority in many parts of the world and will no doubt continue to be in the future. In addition, accurate information on the status of rainforests and their economic, environmental, cultural, and social value will be needed to guide policy makers and others responsible for forest resources. No one expects that the tasks will be easy. But around the world, there are signs that a great variety of approaches are being and will be tried by governments, NGOs, and individuals to save the rainforests of the world.

Works Cited

Cooper, Mark. 1992. "Rain-Forest Crude." *Mother Jones* March/April: 39–47, 75–76.

Cultural Survival. No date. "Ituri Forest Peoples' Fund." http://www. cs.org/specialprojects/ituri/ituri.htm. Accessed February 14, 2001.

de Haan, Cees, Henning Steinfeld, and Harvey Blackburn. 1997. *Livestock and the Environment: Finding a Balance.* Report of a study by the Commission of the European Communities, the World Bank, and the governments of Denmark, France, Germany, the Netherlands, the United Kingdom, and the United States of America coordinated by the Food and Agriculture Organization of the United Nations, the

United States Agency for International Development, and the World Bank. Food and Agriculture Organization website. http://www.fao. org/docrep/x5303e/x5303e00.htm#Contents Accessed February 26, 2001.

Easterbrook, Gregg. 1998a. "Greener Pastures: A New Plan to Save Species." *The New Republic* 2 (March): 14(2).

———. 1998b. "Hot and Not Bothered: The Answer to Global Warming." *The New Republic* 4 (May): 20 (4).

Forests.org, Inc. 2001. "Papua New Guinea: World Bank Ready to Abandon Forsest Reform?" http://forests.org/recent/2001/bankabrd. htm. Accessed February 26, 2001.

Hansen, James E., et al.2000. "Global Warming in the 21st Century: An Alternative Scenario." (An abbreviated version). NASA Goddard Institute for Space Studies website. http://www.giss.nasa.gov/research/impacts/altscenario/. Accessed February 26, 2001.

Hecht, Susanna, and Alexander Cockburn. 1990. *The Fate of the Forest Developers, Destroyers and Defenders of the Amazon.* New York: Harper-Collins.

Kraul, Chris. 1997. "Colombian Tribe Has Oxy Over a Barrel: Indigenous People Vow Mass Suicide if Company Goes after Oil Under Disputed Territory." *Los Angeles Times* (online version), April 25. http://www. amazonwatch.org/newsroom/mediaclips/apr2597lat.html. Accessed February 24, 2001.

Laurance, William F., Mark A. Cochrane, Scott Bergen, Philip M. Fearnside, Patricia Delamônica, Christopher Barber, Sammya D'Angelo, and Tito Fernandes. 2001. "The Future of the Brazilian Amazon." *Science.* 19 (January): 438–439.

Myers, Norman. 1999. "What We Must Do to Counter the Biotic Holocaust." *International Wildlife Magazine* (online version). March–April. http://www.nwf.org/intlwild/1998/contma99.html. Accessed February 24, 2001.

National Wildlife Federation. 2000. "Keep the Wild Alive Campaign— Rosy Periwinkle." http://www.nwf.org/wildalive/periwinkle/. Accessed February 25, 2001.

Rogers, Paul. 2000. "A $5 Billion Plan To Save World's Forests." *San Jose Mercury News* (online version), August 20. http://www.mercurycenter.com/local/center/brazil08200.htm

Ross, John. 2000. "Defending the Forest and Other Crimes." *Sierra* (online version). July–August 2000. http://www.sierraclub.org/sierra/200007/Mexico.asp. Accessed February 24, 2001.

Rowell, Andy, and Peter F. Moore. 2000. *Global Review of Forest Fires.* London: World Wide Fund for Nature.

Sandalow, David B. 2000. "Protecting and Conserving the World's Forests." Address delivered at "Future Directions for Leadership on International Forest Issues" meeting. Washington, DC, National Press Club. 6 (January): Press release. http://www.igc.org/wri/press/frstsandalow.html. Accessed February 24, 2001.

Stevens, William K. 2000. "Global Warming: The Contrarian View." *New York Times* (online version). February 29. http://www.archives.nytimes.com. Accessed August 16, 2000.

U.S. House of Representatives. February 15, 2000. Committee on Resources, Subcommittee on Forests and Forest Health. Testimony of Ron Arnold.http://resourcescommittee.house.gov/106cong/forests/00feb15/ arnold.htm. Accessed February 24, 2001.

Verhovek, Same Howe. 2000. "A Land Use Struggle Over a Forest Bounty." *New York Times*, May 27, 9.

Wilson, Edward O. 1989. "Threats to Biodiversity." *Scientific American*. September. 108–116.

The World Commission on Dams. 2000. "Dams and Development: A New Framework for Decision-Making." http://www.damsreport.org/wcd_overview.htm. Accessed February 26, 2001.

World Rainforest Movement. 2000. "Climate Change Debate Issues." WRM Bulletin # 37:3–10. http://www.wrm.org.uy/index.html. Accessed February 26, 2001.

World Resources Institute. 2000. "Digital Technology Paints Pictures of World's Forest Under Threat." Press release. http://www.wri.org/wri/press/gfw2000–01.html. Accessed February 26, 2001.

World Wildlife Fund. 1997. "1997: The Year the World Caught on Fire." Press release. www.panda.org/forests4life/news/161297_yearfire.cfm

Young, Don. 1998. "The Tongass—A Tale of Two Forests." News release. http://resourcescommittee.house.gov/press/1998/980918tongassoped.htm. Accessed February 26, 2001.

Further Reading

Beckerman, Wilferd. 1996. *Through Green-Colored Glasses: Environmentalism Reconsidered*. Washington, D.C.: Cato Institute.

Bevis, William W. 1995. *Borneo Log: The Struggle for Sarawak's Forests*. Seattle and London: University of Washington Press.

Burger, Julian. 1990. *The Gaia Atlas of First Peoples*. New York: Anchor Books Doubleday.

Goodstein, Eban. 1999. *The Trade-Off Myth: Fact & Fiction About Jobs and the Environment.* Washington, D.C.: Island Press Publishing.

Linden, Eugene. 1999. "Watch What You Eat." *Time* magazine. 31 (December): 1.

Lyall, Sarah. 2000. "A Global Warming Report Predicts Doom for Many Species." *New York Times* 1 (September): electronic version.

Moore, Thomas Gale. 1998. *Climate of Fear: Why We Shouldn't Worry About Global Warming.* Washington, D.C.: Cato Institute.

Pynn, Larry. 2000. *Last Stands: A Journey Through North America's Vanishing Ancient Rainforest.* Corvallis: Oregon State University Press.

Revkin, Andrew C. 2000. "Study Proposes New Strategy to Stem Global Warming." *New York Times.* 19 (August): electronic version.

United Nations Food and Agriculture Organization. 1999. *The State of the World's Forests 1999.*

3

Chronology

Although the process of deforestation worldwide has accelerated dramatically since the middle of the twentieth century, the world's forests have been exploited for centuries—for as long as people have cut trees for fuelwood and have cleared land for agriculture, building, and road construction. During the colonial period in the United States, for example, about 100 million of approximately 850 million acres of original forested land were cut or burned, primarily for agricultural purposes. Between 1850 and 1900, an average of 13.5 square miles of U.S. forest were cleared every day to make room for farms. Forests also were cut for timber supplies used to meet construction and fuel needs, leaving about 580 million acres of forest in the United States by 1900.

As U.S. forests rapidly disappeared, some conservationists began to speak out and write about the dangers of forest loss. This chronology begins with the U.S. conservation movement, which in recent decades has done much to call attention to tropical rainforest destruction worldwide and which since the 1980s has helped spotlight the degradation of temperate rainforests.

The U.S. conservation movement is said to have been inspired by George Perkins Marsh and his book *Man and Nature,* published in 1864. In this book, Marsh pointed out the dangers of deforestation and human interference "with the spontaneous arrangements of the organic or the inorganic world." Marsh warned that the destructive effects of excessive logging in Europe could be replicated in the United States, where vast fertile regions in the East and Midwest were being destroyed at an unprecedented rate.

Americans, like people in other parts of the world, were generally not interested in the preservation of forests and other natural resources. During the Industrial Revolution of the 1800s,

Europeans and Americans were concerned about expanding industry, transportation, agriculture, ranching, and other commercial activities—all of which contributed to deforestation. So conflicts developed between those who wanted to conquer nature and exploit natural resources such as forests and those who wanted to conserve resources or use them wisely.

This chronology reflects the ongoing conflict over the use of natural resources, particularly the timber from all types of forests, including temperate and tropical rainforests. It traces the expanding U.S. conservation efforts that eventually began to help raise awareness of rainforest destruction. The most recent years of the chronology reflect the worldwide expansion of understanding about the need to manage rainforests in a sustainable way, to protect biodiversity and forest products so that they will continue to provide basic needs for people around the world and particularly for those who make the rainforests their home.

1875   Conservationists in the United States establish the American Forestry Association to promote the value of trees and forests.

1892   A rider is added to the U.S. General Revision Act, authorizing the president of the United States to set up "forest reserves on the Public Domain."

        John Muir, an early American conservationist, establishes the Sierra Club to protect the natural habitats of the Sierra Nevada range in the western United States.

1895   Gifford Pinchot, America's leading advocate of forest and other natural resource conservation, becomes head of the Division of Forestry under the Department of Interior.

1900   German-born Frederick Weyerhaeuser travels from the eastern United States to Tacoma, Washington, where he establishes a timber company.

1902   An alliance of forest product manufacturers establishes the National Lumber Manufacturers Association to lobby the U.S. Congress on behalf of the forest products industry and to present the manufacturers' views on forest management.

1905   President Theodore Roosevelt names Gifford Pinchot chief forester of the newly formed U.S. Forest Service, transferred from the Department of the Interior and placed under the direction of the Department of Agriculture. Pinchot promises to "advocate nothing in the way of forestry that will not pay"—that is, to manage the forests so they will be self-supporting.

1916   The National Park Service is established in the United States.

1930   The U.S. Congress passes and President Herbert Hoover signs the Knutson-Vandenberg Act, which is designed to restore forest areas. The act gives the U.S. Forest Service the authority to require anyone buying timber from national forests to deposit with the federal treasury enough funds to pay for the reforestation of the cut area.

1936   U.S. President Franklin Roosevelt convenes the first North American Wildlife Conference. The conference creates the General Wildlife Federation, a private organization designed to restore and preserve the continent's disappearing wildlife resources.

1944   Representatives of forty-four nations attending a global conference establish the World Bank, which is able to invest millions of dollars annually in development projects in nonindustrialized nations.

        Because of dwindling supplies of timber on private lands and increasing demands for wood products nationwide, the U.S. Congress passes the Sustained Yield Management Act, allowing the Forest Service to sell timber from national forest areas that can be renewed, that is, reforested, within a specified period of time.

1946   The U.S. Bureau of Land Management is created to oversee public lands and maintain records of these lands.

1948   Leaders of national governments and nongovernmental organizations (NGOs) meet in Fontainebleau, France, to form the International Union for the Protection of Nature, which later will be renamed the International

Union for Conservation of Nature and Natural Resources, and is now the World Conservation Union. The organization will become an alliance of government agencies and NGOs, representing 120 nations, that try to find solutions to environmental problems and work to conserve natural resources.

1949   Aldo Leopold's *A Sand Country Almanac* is published posthumously. The book presents Leopold's views on conservation in an essay titled "The Land Ethic," which declares that conservation is "a state of harmony between men and land." Leopold also maintains that people should respect, admire, and love the order and integrity of natural communities—now called *ecosystems*.

1950   At this time, 15 percent of the earth's land surface is covered with rainforest, and the rate of forest destruction is accelerating.

1951   The Nature Conservancy is organized to preserve biological diversity by setting aside large tracts of natural land.

1960   In response to increasing criticism of U.S. Forest Service management practices, the U.S. Congress passes and President Dwight Eisenhower signs the Multiple Use–Sustained Yield Act. The federal law requires the U.S. Forest Service to manage the renewable resources of the national forests so that "they are utilized in the combination that will best meet the needs of the American people" and so that there will be "high-level annual or regular periodic output of the various renewable resources of the national forests without impairment of the productivity of the land."

1961   Stewart Udall, U.S. secretary of the interior, articulates his views on the need for stewardship of earth's resources, including forests.

1963   The U.S. Forest Service acquires Grey Towers, the former home of Gifford Pinchot, first chief of the Forest Service. The French chateau-style mansion in Milford, Pennsylvania, is surrounded by 102 acres of forest, formal gardens, and meadows.

1964   After sixty-six revisions, the U.S. Congress passes and President Lyndon Johnson signs the Wilderness Act. The act creates a system for preserving wilderness areas, including temperate and tropical rainforests under U.S. jurisdiction.

1965   For the first time, people from industrialized nations make contact with the Kayapo, an indigenous group in the Amazon who call themselves "people of the water's source." Soon thereafter, their way of life begins to be threatened by invasions of loggers and others wanting to exploit the rainforest.

The United Nations sponsors a World Population Conference, calling attention to the link between rapid global population growth and environmental problems such as pollution and rainforest destruction.

1968   *The Population Bomb,* by noted Stanford University professor Paul Ehrlich, is published. Ehrlich warns of global ecosystem deterioration, famine, and other major problems resulting from exploding populations.

In Paris, representatives of international organizations meet in a conference called the Intergovernmental Conference of Experts on the Scientific Basis for Rational Use and Conservation of the Resources of the Biosphere. The conference marks the beginning of a major international focus on environmental concerns.

A detailed plan to set up a protected park for one group of indigenous people, the Yanomami, is presented to the Brazilian government, but the plan is ignored.

1969   The U.S. Congress passes the National Environmental Policy Act, which incorporates some of the standards of environmental ethics established by Aldo Leopold years earlier.

Because of concerns about the destruction of South America's indigenous communities and their forest land, Survival International is organized in Great Britain to campaign for the rights of indigenous people.

1970   In Brazil, the government begins a colonization program known as Polonoroeste in largely inaccessible Rondônia and Mato Grosso in southern Brazil. Since the program is designed to resettle people from overpopulated areas of Brazil, a major highway, BR-364, is cut into the rainforest.

The U.S. Congress creates the Environmental Protection Agency, which is charged with enforcing national standards for environmental quality.

U.S. political activists concerned about environmental issues establish the League of Conservation Voters. Their purpose is to campaign for, elect, and support public officials at all levels of government who sponsor or vote for legislation that protects the environment.

The first celebration of Earth Day, an idea developed by then U.S. Senator Gaylord Nelson, takes place on April 22 in cities, towns, and villages across the United States. At least 20 million people participate, setting the stage for the modern environmental movement in the United States, which incorporates rainforest preservation as a major part of its agenda.

1971   The U.S. Congress passes the Alaska Native Claims Settlement Act. The act, which affects about 1.7 million acres of temperate rainforest in the Tongass and Chugach National Forests in Alaska, allows twelve regional Alaska Native corporations to select acreage that can be logged or mined.

1972   Representatives from 113 nations meet in Stockholm, Sweden, for a UN Conference on the Human Environment to discuss global environmental concerns. World Environment Day is established by the UN General Assembly, and its first celebration marks the opening of the conference. Another resolution, adopted by the General Assembly the same day, leads to the creation of the UN Environment Programme to encourage and coordinate environmental programs within a variety of UN agencies.

1973   Congress passes the Endangered Species Act designed to prevent the loss of species. One provision of the law requires that the secretary of the Department of Interior, in cooperation with the secretary of the Department of Commerce, periodically prepare lists of endangered and threatened species for publication in the *Federal Register.* The law also bans the export and killing of any animal or plant listed as endangered. No endangered species can be bought, sold, or taken from the wild without a special permit. Other provisions require protection for the habitats of endangered and threatened species, many of which are in rainforest areas.

1974   The U.S. Congress passes and President Gerald Ford signs the Sikes Act. One of the provisions of the act requires federal agencies, such as the Forest Service and the Bureau of Land Management, to protect the habitats of endangered species and to initiate efforts to conserve fish and wildlife.

The Convention on International Trade in Endangered Species of Wild Fauna and Flora becomes an international law. The agreement is an attempt to control the illegal international trade in the products of endangered species (such as ivory and fur).

1976   The U.S. Congress passes the National Forest Management Act, which requires each national forest to prepare a management plan and sets limits on the size of an area that can be harvested by clear-cutting methods. The law allows timber cutting only on "such lands that can be restocked within five years after harvest."

1979   James E. Lovelock's *Gaia: A New Look at Life on Earth* is published. Using the name that ancient Greeks gave to the goddess of the Earth, Gaia, the book presents a theory about how the earth works—as a single living system of which humankind is only one small part. Lovelock argues as well that just as life-forms adapt to the Earth, so too does the planet adapt to life-forms, regulating Earth's climate and maintaining stability in spite of environmental changes. According to the Gaia theory, exploitive human activities, such as massive deforestation and fossil fuel use, are creating imbalances that could destroy life as we know it but that would allow another form of life to emerge.

1980   A report on global environmental trends, commissioned
        by President Jimmy Carter in 1977, is published under
        the title *"The Global 2000 Report to the President."* The re-
        port describes serious environmental stresses worldwide
        and predicts that "life for most people on Earth will be
        more precarious in 2000 . . . unless the nations of the
        world act decisively to alter current trends."

        Thomas E. Lovejoy, a nationally known biologist, sug-
        gests an idea for rainforest preservation in nonindustri-
        alized countries: swapping debt for nature-conservation.

        The U.S. Congress passes the Alaska National Interest
        Lands Conservation Act, which sets aside huge areas of
        federal land, including rainforest areas in Alaska, for
        preservation, recreation, and resource management.

1982   The UN Food and Agriculture Organization publishes a
        study of worldwide deforestation, reporting a tropical
        rainforest loss of 11 million hectares (27 million acres) an-
        nually. The report reveals that some reforestation is tak-
        ing place, but there is still a net loss of 10 million hectares
        (25 million acres) each year.

1983   James Watt, appointed by President Ronald Reagan, re-
        signs as U.S. secretary of the interior, primarily because
        of negative public reaction to his decisions to allow
        drilling for oil and gas and mining on public lands, in-
        cluding protected forest lands.

        The United Nations establishes the World Commission
        on Environment and Development headed by Gro
        Harlem Brundtland, leader of the Norwegian Labor
        Party. She leads twenty-two policy makers and scientists
        from as many countries in developing proposals for
        dealing with crucial problems in environment conserva-
        tion and development and to find ways to foster interna-
        tional cooperation.

1985   Rubber tappers from the Brazilian states of Rondônia,
        Acre, Amazonas, and Pará meet for the first time in Brasília
        to discuss ways to preserve their forest homes and liveli-
        hood; they issue a manifesto on Amazonian development.

1987 After three years of research on global development issues, the World Commission on Environment and Development publishes "The Brundtland Report," which
outlines a strategy for global sustainable development—
that is, development that does not destroy ecological systems. Among its seven goals is "merging environment
and economics in decision-making" and assessing "potential impacts of new technologies before they are
widely used, in order that their production, use and disposal do not overstress environmental resources."

Members of rainforest activist groups from around the
world meet in Malaysia to organize a global movement
to help save the world's rainforests. They also take part
in the first annual Rainforest Week, which is designed to
focus attention on the need to stop the destruction of
tropical rainforests.

1988 James E. Hansen, a scientist with the NASA Goddard Institute for Space Studies, tells a U.S. Senate committee
that he is 99 percent certain that global warming is under
way due to the buildup of greenhouse gases, which is
partly caused by deforestation and the release of carbon
dioxide (CO_2) from fallen trees but is primarily the result
of the massive fossil fuel use.

In December, Chico Mendes of Amazonia, who has organized rubber tappers working to preserve their way of
life, is murdered in the state of Acre, Brazil.

1989 Supported by major U.S. corporations, representatives of
250 U.S. groups, such as the American Mining Congress,
the National Cattlemen's Association, and the Western
Environmental Trade Association, meet in Bellevue,
Washington, to form the "wise use movement" and develop strategies to fight environmentalism. Among the
coalition's agenda are anti-environmental campaigns de-
manding the freedom to clear-cut ancient forests, open up
national parks and wilderness areas to mining and energy
production, and dismantle the Endangered Species Act.

In April, a delegation that includes three Amazonian indigenous leaders travels from Brazil to London, En-

gland, to talk to the Prince of Wales and other influential people about their plan to extend and protect a Brazilian national park and Native reserve in the Amazon.

In November, then Senator Albert Gore and fourteen of his colleagues in the U.S. Congress meet with members of the European Parliament to form Global Legislators Organized for a Balanced Environment (GLOBE). The legislators make plans to meet twice annually to exchange information on global environmental problems, including deforestation, and to discuss possible solutions.

1990   The government of Belize and LightHawk, a U.S. environmental group, announce the creation of the Bladen National Park, the first national park established in this small Central American nation. The park will protect more than 80,000 acres (197,600 hectares) of virgin tropical forest.

Conservation International launches its Rain Forest Imperative, a ten-year strategy to sustain the richest and most critically threatened forests on Earth. The organization's plan calls attention to areas needing immediate protection, including fifteen tropical forest "hot spots," three major tropical wilderness areas, and the two largest regions of at-risk temperate rainforest.

The U.S. Congress passes the Tongass Timber Reform Act, which is designed to eliminate the mandatory cutting of timber to meet targets set by the U.S. Forest Service. Instead, only enough timber to meet realistic market demands will be cut in this temperate rainforest.

In July, more than 1 million Ecuadoran indigenous people stage a week-long nonviolent protest, temporarily stopping expanded oil drilling on a rainforest reserve. But the government retaliates by arresting and jailing many protesters, and oil exploration continues.

The northern spotted owl, whose habitat is ancient forests of the Pacific Northwest in the United States, is listed as an endangered species.

1991   In April, a group of North American activists begins a two-month educational and cultural tour called a *chautauqua* to raise awareness and educate the public on the ecology of temperate rainforests. The chautauqua starts in British Columbia, Canada, travels along the Pacific Northwest coast, and ends in San Diego, California.

During the UN Environment Day in June, Claes Nobel, great-grandnephew of Alfred Nobel, who initiated the prestigious Nobel prizes, awards the first Nobel prizes for environmental achievement in seven categories, including preservation of nature and natural resources.

In September, the tenth World Forest Congress convenes in Paris, a collaborative effort of France and the UN Food and Agriculture Organization. The purpose of the congress is to raise global public awareness of forest issues.

On December 6, the government of Belize signs a statute creating the Chiquibul National Park to protect lowland tropical forest areas along the border with Guatemala.

Colombia adopts a new constitution, which requires that all natural resource development projects include a formal consultation process with indigenous communities.

1992   To protect sources of an anticancer chemical found in the bark of yew trees, the U.S. Congress passes the Pacific Yew Act. The act is designed to conserve yews in ancient forest stands, some of which are part of coastal rainforests.

In January, a new coalition of sixty-two NGOs in the Amazon forms to work toward preservation of Brazil's tropical forests.

British Columbia's New Democratic Party government announces in late January that it will ban logging for eighteen months in the temperate forests of Tsitika, Tashish, and Walbran Valleys and that it will freeze logging operations in three more areas in the province's interior.

In the United States, the Natural Resources Defense Council, a nonprofit environmental organization, cosponsors with the Yale Law School the first major international conference on the relationship between human rights and environmental protection.

In June, the Earth Summit convenes in Rio de Janeiro, Brazil.

The Brazilian government sets aside rainforest land for indigenous peoples, the Yanomami and the Kayapo.

1993   The United Nations declares this year as the International Year of the World's Indigenous Peoples, in which policy makers and an informed public worldwide attempt to set up guidelines to establish land rights for indigenous people and to resolve conflicts over natural resources.

Rallies and blockades are conducted to demonstrate against logging in Clayoquot Sound, British Columbia, Canada.

1995   The UN Commission on Sustainable Development establishes the Intergovernmental Panel on Forests to identify proposals for protecting and conserving forests.

Twelve nations with 90 percent of the world's temperate and boreal forests come together as the Montreal Process Working Group to endorse a set of criteria for sustainable forest management.

1997   The Nature Conservancy, the American Electric Power Company, and the Bolivian government provide more than $11 million to expand the Noel Kempff National Park in Bolivia. The expansion furthers efforts to reduce carbon emissions and protects the ecologically rich area.

More than 160 nations meet in Kyoto, Japan, to negotiate binding limitations on greenhouse gases for the developed nations. The meeting results in the Kyoto Protocol, in which developed nations agree to limit their greenhouse gas emissions, relative to the levels emitted in 1990.

1998   The U.S. Congress passes the Tropical Forest Conservation Act. The act authorizes reduction of the debt that countries with tropical forests owe to the United States; in exchange, these countries must initiate conservation measures to preserve those forests.

Leaders from the G-8 (the Group of Eight industrialized nations: Britain, Canada, France, Germany, Italy, Japan, Russia, and the United States) adopt an Action Program on Forests that combats illegal logging and calls on nations to engage NGOs and the private sector in programs to protect forests.

Citizens of fourteen countries around the world gather in Montevideo, Uruguay, because of their urgent concerns over the accelerating invasion of millions of hectares of land and forests by pulpwood, oil palm, rubber, and other industrial tree plantations. By signing the Montevideo Declaration, the group members pledge to join an international campaign to support local peoples' rights and struggles against the invasion of their lands by these plantations; to encourage awareness of the negative social and environmental impacts of large-scale tree plantations; and to change the conditions that make such plantations possible.

1999   In Squamish, British Columbia, a group of women begin a blockade of the logging roads leading to the contested Stoltmann Wilderness in the Elaho Valley north of Vancouver.

2000   The World Resources Institute launches Global Forest Watch, a tracking system using satellite imagery and ground-level observations to improve knowledge about the state of forests worldwide.

A 1,900-square-mile reserve is established in Gabon, Africa.

A joint initiative of Conservation International, the World Bank, and the Global Environment Facility creates a new, multimillion-dollar global fund aimed at slowing species loss by focusing on the hot spot regions.

The latest edition of World Resources Institute's annual report on the world's resources, entitled *People and Ecosystems: The Fraying Web of Life*, is presented to the world's environment ministers in Bergen, Norway, on September 15. Prepared by the UN Development Programme, the UN Environment Programme, the World Bank, and the World Resources Institute, the Millennial Edition of the report presents a comprehensive assessment of five of the world's major ecosystems, including forest systems.

2001   For the first time since the 1700s, scientists declare a primate extinct: the West African monkey called Miss Waldron's red colobus, a member of the taxonomic group to which humans belong; the species once lived in the rainforests of Ghana and the Ivory Coast.

4

People and Events

Countless people and events are involved in the protection and exploitation of rainforests: the people who are associated with rainforest preservation on the one side and those who advocate for timber and other industries that use or abuse rainforest resources on the other. Some of the conflicts between these groups have been described in Chapter 2. The profiles in this chapter are not meant to be comprehensive. Only a few of the vast number of individuals and events worldwide that have directly or indirectly contributed to rainforest preservation or destruction are highlighted. Certainly some important people and events are missing. But the representative profiles illustrate the diversity of activities focusing on tropical and temperate rainforests, climate change, and species loss attributed to deforestation around the globe. Included are descriptions of present and past grassroots activists as well as leaders of environmental groups, government officials, and business executives.

Abbey, Edward (1927–1989)

Called a radical environmentalist, writer Edward Abbey is today best known for his 1974 novel *The Monkey-Wrench Gang*, which features characters who use wrenches and other tools to damage construction equipment and other machinery that destroy wilderness areas. Abbey's novel was a motivating factor in the 1980s Earth First! movement—which conducted its own monkey-wrenching activities to protect the ancient forests that make up much of the coastal temperate rainforests in the Pacific Northwest.

Abbey was born in Home, Pennsylvania, and during the late 1940s and early 1950s he was educated at the University of New Mexico and the University of Edinburgh in Scotland. His fifteen-year tenure as a park ranger in the American Southwest inspired some of his writings. Besides his 1974 novel, Abbey wrote other works on the environment, including *Fire on the Mountain* (1962), *Desert Solitaire* (1968), and *Beyond the Wall* (1984).

Further Reading: Abbey, Edward. 1988. Reprint. *Desert Solitaire.* Tucson: University of Arizona Press. Original edition: New York: McGraw Hill, 1968; McCann, Garth. 1997. *Edward Abbey.* Boise, ID: Boise State University.

Alaska Native Claims Settlement Act (1971)

To settle the long-standing land claims of Native Alaskan Inuits, Aleuts, and Indians, the U.S. Congress passed the Alaska Native Claims Settlement Act (ANCSA) in 1971. The act established 12 regional corporations and more than 200 village corporations that were awarded title to 44 million acres of federal land and paid $962.5 million. Later amendments to the law were passed to protect long-term Native American control of the land.

Since ANCSA became law, Native corporations have developed logging as well as mining and oil drilling operations on various land sites. About 1.7 million acres of temperate rainforest in the Tongass and Chugach National Forests in Alaska are affected by the ANCSA. Although these lands are in national forests, they are now privately owned and are exempt from federal laws protecting public resources. More than half the timber on Alaska Native corporate land has been logged and the remainder could be depleted by 2010.

Further Reading: Berger, Thomas R. 1985. *Village Journey: The Report of the Alaska Native Review Commission.* New York: Hill & Wang; Chance, Norman A. 1990. *The Inupiat and Arctic Alaska: An Ethnography of Development.* New York: Holt, Rinehart & Winston; Pynn, Larry. 2000. *Last Stands: A Journey Through North America's Vanishing Ancient Rainforest.* Corvallis: Oregon State University Press.

Blythe, Stephen (1952–)

A family physician and nutritionist, Stephen Blythe has made numerous trips to rainforests to study plant medicines. He has also provided health care and supervised the training of community health workers in a Guatemalan refugee camp in the jungles of southern Mexico. In addition, he has participated in conferences in Belize and Costa Rica that focused on rainforest plants containing pharmaceutical ingredients.

Born in Port Arthur, Texas, Blythe completed a graduate degree in public health and nutrition at Pennsylvania State University (1976–1978), which included three months of fieldwork with the White Mountain Apache in Whiteriver, Arizona. From 1978 to 1979, he conducted clinical nutrition research at the New England Deaconess Hospital in Boston and spent five months volunteering his services to Oxfam-America, a private voluntary relief organization, where he assisted in the preparation of a book analyzing the system of food aid delivery to developing nations. In 1983, he received his doctor of osteopathy (D.O.) degree from the University of New England College of Osteopathic Medicine in Biddeford, Maine. He has practiced medicine in Colorado and currently practices in Florida, where he lives with his wife and two daughters.

Blythe's interest in travel led to the development of the Travel Health Information Service, an Internet site that received a number of awards. In 2000, the site was purchased by MedicinePlanet.com, a new Internet venture, for which Blythe serves as a consultant. In addition, he is developing a rainforest education site, RainforestEducation.com, which includes information on medicines from the rainforest and threats to rainforests.

Brown, Lester R. (1934–)

Lester R. Brown founded and is chairman of the board of the internationally acclaimed Worldwatch Institute, which tracks environmental trends. Brown is known as an influential thinker and the guru of the global environmental movement. He began his career as a farmer, growing tomatoes in New Jersey while in high school and college. Brown earned graduate degrees in agricultural economics at the University of Maryland and Harvard, and from 1959 to 1969, he served as administrator of the U.S. Department of Agriculture's International Agricultural Development Service, traveling extensively worldwide. He spent most of his travel time in rural areas abroad, where he observed the effects of environmental abuse—destruction of rainforests, overgrazing, soil erosion, desertification, and general deterioration of the agricultural base of many countries—conditions that he recognized would undermine these countries' economies.

As the result of such observations, Brown began to monitor trends in environmental damage and founded Worldwatch Institute in 1974 with a grant from the Rockefeller Brothers Fund. Brown and the institute have received numerous international environmental awards and prizes, and his *State of the World* reports, published annually by the institute, have been translated into at least thirty languages and are widely respected by the global environmental movement. Other publications that Brown has launched include *World Watch*, a bimonthly magazine featuring articles on the institute's research, the *Environmental Alert* book series, and the annual publication, *Vital Signs: The Trends That Are Shaping Our Future*. Brown has also authored numerous other books on environmental and population topics.

Brown speaks to groups around the world and counsels government leaders on policy changes that will help protect the planet, calling for more fuel-efficient forms of transportation, much more recycling of goods made from nonrenewable resources, and most importantly, a restructuring of the global economy so that the environment is sustained. He believes that environmental deterioration and social disintegration are linked, that national security itself requires the conservation and wise management of the nation's and the world's natural resource base—an important part of which are the world's rainforests.

Further Reading: Brown, Lester R., Gary Gardner, and Brian Halweil. 1999. *Beyond Malthus: Nineteen Dimensions of the Population Challenge.* New York: W.W. Norton & Company; Brown, Lester R., Linda Starke, ed., and Brian Halweil. 2000. *Vital Signs 2000: The Environment Trends That Are Shaping Our Future.* New York: W.W. Norton & Company; Brown, Lester R., Christopher Flavin (contributor), and Hilary French (contributor). 2000. *State of the World 2000.* New York: W.W. Norton & Company.

Earth Day (1970–)

The first Earth Day, in 1970, was based on an idea conceived by a U.S. senator from Wisconsin, Gaylord Nelson, who became counselor of The Wilderness Society in 1981, and was honorary cochair of Earth Day on its twentieth anniversary in 1990. According to Senator Nelson,

> the idea for Earth Day evolved over a period of seven years, starting in 1962. For several years, it had been troubling me that the state of our environment was simply a nonissue in the politics of the country. Finally, in November 1962, an idea occurred to me that was, I thought, a

virtual cinch to put the environment into the political "limelight" once and for all. The idea was to persuade President Kennedy to give visibility to this issue by going on a national conservation tour. I flew to Washington to discuss the proposal with Attorney General Robert Kennedy, who liked the idea. So did the President. The President began his five-day, eleven-state conservation tour in September 1963. For many reasons the tour did not succeed in putting the issue onto the national political agenda. However, it was the germ of the idea that ultimately flowered into Earth Day (Nelson 2000).

The first Earth Day event in 1970 was on April 22, the day celebrated annually from then on. It was coordinated by Denis Hayes, who twenty years later chaired the first International Earth Day, which enlisted 200 million participants in 141 countries. In 1999, organizers of the original Earth Day established the Earth Day Network with headquarters in Seattle, Washington. The network maintains a website to help coordinate and organize annual events and campaigns that promote a healthy and sustainable environment.

Wherever Earth Day is celebrated around the globe, events may include marches, seminars, fairs, or other ceremonies that focus on preserving natural resources. Numerous events focus on efforts to protect rainforests. For example, on Earth Day 2000 the U'wa tribe, which has been fighting against oil drilling in the northern Cocuy rainforest of Colombia, held a traditional indigenous ceremony for the fertility of Mother Earth. During the ceremony, tribal members wearing traditional dress sang and danced, asking the U'wa god to care for the Earth and allow it to continue to conceive and reproduce its beauty, despite the troubles surrounding the U'wa people. In Gabon, Africa, a network of traditional talking drums sent messages urging preservation of the Ipassa-Mingouli Forest, which along with its indigenous people, is being threatened by excessive logging. Travelling from the forest to the port city of Libreville, the messages were webcast on Earth Day as part of an effort to preserve the forest by petitioning for methods of sustainable logging and establishing collaboration between loggers, conservationists, and the local government.

Further Reading: "Earth Day 2000." 2000. Time.com (special online edition). http://www.time.com/time/reports/earthday2000/assessment. html. Accessed February 27, 2001; "Earthday Network." 2000. Earth Day website. http://www.earthday.net/. Accessed February 27, 2001; Nelson, Gaylord. 2000. "All About Earth Day." The Wilderness Society website. http://earthday.wilderness.org/history/index.htm. Accessed February 27, 2001; Shulman, Jeffrey, Teresa Rogers, and Larry Raymond. 1992. *Gaylord Nelson: A Day for the Earth*. New York: 21st Century Books; Stanley, Phyllis M. 1996. *American Environmental Heroes*. Berkeley Heights, NJ: Enslow Publishers.

Ehrlich, Paul R. (1932–) and Anne H. (1933–)

Stanford University professors Paul and Anne Ehrlich are well-known worldwide as population experts and advocates for population control. Paul Ehrlich is the author of hundreds of scientific papers and magazine articles and dozens of books, mainly on ecology and evolution and the value of biodiversity. He is widely known for his book *The Population Bomb*, published in 1968. He is honorary president of Zero Population Growth, president of the American Institute of Biological Sciences, a fellow of the American Association for the Advancement of Science and the American Academy of Arts and Sciences, and a member of the U.S. National Academy of Sciences and the American Philosophical Society. Educated in the public schools of Pennsylvania and New Jersey, Paul Ehrlich received a graduate degree from the University of Pennsylvania in 1955. He earned a Ph.D. in biology at the University of Kansas in 1957 and has done extensive field, laboratory, and theoretical research in the field of population biology, which includes ecology, evolutionary biology, and behavior studies. He also collaborates with his wife, Anne Ehrlich, in policy research on human ecology.

Ehrlich's research has taken him to all continents, and his work is known worldwide, often serving as a resource for people working on rainforest preservation projects. He has received numerous awards, has appeared frequently on television and radio programs, and has given hundreds of public lectures on ecology, human ecology, and evolution—consistently promoting ways that governments and individuals can make changes that will help sustain life on the planet.

Anne Ehrlich has written or coauthored with her husband numerous articles and books on population biology and has written extensively on such public issues as population control and how population growth affects the environment, particularly ecosystems such as rainforests. She is a consultant for many rainforest action groups. Born in Iowa, Anne Ehrlich graduated from the University of Kansas in 1955. She began her career in the Department of Entomology at the University of Kansas, and in 1959 she joined the research staff of biological sciences at Stanford University, becoming associate director for the Center for Conservation Biology in 1987 and policy coordinator in 1992.

Anne and Paul Ehrlich frequently have pointed out in their writings and in interviews that the rapid increase in human population is already having a deleterious effect on the environment and the quality of human life. The Ehrlichs warn that no nation can be secure if overpopulation leads to degradation and destruction of ecosystems. Soil fertility, plant pollination, nutrient recycling, a balance of gases in the atmosphere, biodiversity, and the hydrologic cycle are all vital to life on Earth. According to the Ehrlichs, "the United States can be considered (in terms of its impact) the world's most overpopulated nation," because "the average American is a superconsumer, and the nation generally uses inefficient, environmentally damaging technologies." On a per capita basis, the United States thus poses a much greater threat to the planet than does a poor nation that exploits its resources.

Further Reading: Ehrlich, Paul R. 2000. *Human Natures: Genes, Cultures, and the Human Prospect.* Washington, DC: Island Press; Ehrlich, Paul R., and Anne H. Ehrlich. 1990. *The Population Explosion.* New York: Simon & Schuster; Ehrlich, Paul R., and Anne H. Ehrlich. 1991a. "Population Growth and Environmental Security," *Georgia Review.* (Summer): 223–232; Ehrlich, Paul R., and Anne H. Ehrlich. 1991b. *Healing the Planet.* New York: Addison-Wesley; Ehrlich, Paul R., and Anne H. Ehrlich. 1996. *Betrayal of Science and Reason: How Anti-Environmental Rhetoric Threatens Our Future.* Washington, DC: Island Press; Motavalli, Jim. 1996. "Conversations: Paul and Anne Ehrlich." Emagazine.com (online version of *E/The Environmental Magazine*). November-December: http://www.emagazine.com/november-december_1996/1196conv.html. Accessed February 2, 2001.

Goldman Environmental Prize (1990–)

Richard and Rhoda Goldman established the Goldman Environmental Foundation in 1990 and offered an annual prize to honor six grassroots "environmental heroes" from each of the inhabited continental regions. Each year, a network of internationally known environmental organizations and a confidential panel of environmental experts select recipients based on their sustained and important environmental achievements. The prize is a way to honor and recognize environmental heroes for their efforts.

Among the first prize winners in 1990 was Harrison Ngau, a Kayan tribesman in Malaysia. Ngau was harassed by the Malaysian government and imprisoned for his efforts to help the Penan, indigenous people in the state of Sarawak, in their attempts to halt the rampant logging that has been destroying their way of life and some of earth's most ancient tropical forests.

Over the years, other recipients have been rewarded for their work to preserve and protect rainforests. One winner in 2000 was Nat Quansah, an ethnobotanist who teaches traditional medicine at a Madagascar University, where he reawakened an interest in herbal medicines because of their "efficacy and economics." Quansah explains: "The plants work effectively to heal, and because the majority of people are poor and can't afford western medical care, they are looking for alternatives." A news story reports that Quansah believes medicinal flora give local villagers a reason to save the forests that harbor the plants. "It provides a tangible incentive to preserve nature," he said. "The forests help us, and we help the forests. It's a balance" (Martin 2000). Quansah will use his prize to reestablish a village clinic and medicinal herb garden that he had to close because of a lack of funds.

Another winner in 2000 was Rodolfo Montiel Flores (see sketch in this chapter), who was jailed in 1999 for his efforts to prevent the destruction of his homeland. He organized peasants to block roads after loggers began felling virgin forests near his village in the mountains north of Acapulco, Mexico.

Further Reading: Holtz, Debra Levi. 2000. "Goldman Winners Join to Aid Activists—Watchdog Group Formed." *San Francisco Chronicle,* July 14; "Goldman Environmental Prize" website. http://www.goldmanprize. org/index.html; Martin, Glen. 2000. "7 Environmental Heroes Honored." *San Francisco Chronicle,* 17 April. A2.

Kyoto Protocol (1997)

In 1997, more than 160 nations met in Kyoto, Japan, to negotiate limitations on greenhouse gas emissions, which have long been linked to global warming. Many scientists theorize that extensive logging and large-scale burning of rainforests contribute to the release of large amounts of carbon dioxide (CO_2), which interacts with other chemicals in the atmosphere, creating a buildup of greenhouse gases. The Kyoto meeting resulted in the Kyoto Protocol, in which the developed nations agreed to limit their greenhouse gas emissions to 1990 levels. The six gases covered by the protocol are carbon dioxide (CO_2), methane (CH_4), nitrous oxide (N_2O), hydrofluorocarbons (HFCs), perfluorocarbons (PFCs), and sulphur hexafluoride (SF_6). The most prominent of these and the most pervasive in human economic activity is CO_2, produced when wood or fossil fuels such as oil, coal, and gas are burned.

A central feature of the treaty is a set of binding emissions targets for developed nations. The specific limits vary from country to country, but the limits for key industrial powers of the European Union, Japan, and the United States are similar—8 percent below 1990 emissions levels for the European Union, 7 percent for the United States, and 6 percent for Japan. The protocol calls for emissions targets to be reached over a five-year budget period, which allows emissions to be averaged over five years. This increases flexibility by helping to smooth out short-term fluctuations in economic performance or weather, either of which could spike emissions in a particular year, according to the U.S. Bureau of Oceans and International Environmental and Scientific Affairs (BOIESA).

The first budget period will be 2008–2012, which allows enough time for U.S. companies to make the transition to greater energy efficiency and lower-carbon technologies. Activities that absorb carbon, such as planting trees to create "carbon sinks," will be offset against emissions targets. At Kyoto, the United States insisted that the carbon sinks be allowed in order to encourage reforestation. As the BOIESA stated: "Accounting for the role of forests is critical to a comprehensive and environmentally responsible approach to climate change. It also provides the private sector with low-cost opportunities to reduce emissions."

The treaty also includes an emissions trading regime, whereby countries or companies can purchase less expensive emissions permits from countries that have more permits than they need because they have already met their targets with room to spare. Structured effectively, emissions trading can provide a powerful economic incentive to cut emissions while also allowing important flexibility for taking cost-effective actions, BOIESA contends.

The Kyoto Protocol has been controversial, however. As an article in *Foreign Affairs* points out,

> those who support the treaty are pleased that richer nations participating in the protocol agreed to cut their collective emissions of the greenhouse gases that cause global warming to about 5 percent less than 1990 levels. . . . These optimists also applaud features of the Kyoto accord

designed to hold down the costs of achieving these reductions. . . . [Opponents] see Kyoto as a costly defeat. They note that there is no solid proof that human-induced climate change will occur or that its adverse effects would be serious were it to happen. At the same time, the expense of reducing greenhouse gas emissions to meet the Kyoto targets will be substantial, and pessimists believe that the effort will make participating countries less competitive" (Jacoby, Prinn, and Schmalensee 1998).

Although 100 nations had signed the Kyoto Protocol by 2000, none of the industrialized countries had ratified the agreement. Before it can go into effect, the treaty must be ratified by at least 55 countries, accounting for at least 55 percent of the total 1990 carbon dioxide emissions of developed countries. U.S. ratification requires the advice and consent of the Senate.

Further Reading: Fletcher, Susan R. March 6, 2000. "Global Climate Change Treaty: The Kyoto Protocol." Congressional Research Service Issue Brief for Congress, #98–2. National Council for Science and the Environment website. http://www.cnie.org/nle/clim–3.html. Accessed January 31, 2001; Jacoby, Henry D., Ronald G. Prinn, and Richard Schmalensee. 1998. "Kyoto's Unfinished Business." *Foreign Affairs* (online version) July-August. http://web.mit.edu/globalchange/www/rpt32.html. Accessed February 27, 2001; Morrissey, Wayne A., and John R. Justus. July 17, 2000. "Global Climate Change." Congressional Research Service Issue Brief for Congress, #IB89005. National Council for Science and the Environment website. http://www.cnie.org/nle/clim–2.html. Accessed February 1, 2001; United States Department of State, Bureau of Oceans and International Environmental and Scientific Affairs. 1998. "The Kyoto Protocol on Climate Change Fact Sheet." http://www.state.gov/www/global/oes/fs_kyoto_climate_980115.html. Accessed March 5, 2001.

Lovejoy, Thomas Eugene (1941–)

A biologist, conservationist, and leading Amazon researcher, Thomas Lovejoy is counselor to the secretary of environmental affairs of the Smithsonian Institution, Washington, D.C., and chief biodiversity advisor for the World Bank. He is considered one of the world's leading environmentalists and serves on the boards of a variety of wildlife conservation organizations. He has also been an administrator of numerous conservation and research institutions.

Educated at Yale University, Lovejoy earned a Ph.D. in biology in 1971. During most of his career, he has called for action to protect the rapidly disappearing tropical rainforests and the species that are destroyed with them. In the early 1980s, he articulated a new idea for rainforest preservation: debt-for-nature swaps. Following such a strategy, an environmental group loans money to a tropical nation so it can repay its debt to international development banks. In exchange, the debtor nation pays back its debt to the environmental group, which uses the interest on the loan to buy tracts of rainforest land or to pay for local projects that would preserve forests. In the United States, legislation was passed in 1987 allowing various environmental groups to act upon Lovejoy's idea.

In 1980, Lovejoy was the first person to use the term "biological diversity," and he made the first projection of global extinction rates in the *Global 2000 Report to the President* that same year. He also founded the public television series *Nature*, and for many years served as principal advisor to the series.

Further Reading: Laurence, William F. 1998. "Fragments of the Forest." *Natural History.* July-August: 35–38; Lovejoy, Thomas. 1985. *Magnificent Exception.* Washington, DC: Smithsonian Institution; Laurence, William F. 1992. "Voyage to Diversity." *Bioscience.* November: 654–655; Laurence, William F. "Rain Forest Devastation is Real." *Wall Street Journal,* July 17: A35.

Marsh, George Perkins (1801–1882)

Architect, linguist, politician, diplomat, geographer, scholar— these are just a few of the terms used to describe George Perkins Marsh, called the "original environmentalist" by some conservationists today. Marsh was an expert in diverse fields, but of his many accomplishments, the one most recognized is his scholarly book *Man and Nature: Or, Physical Geography as Modified by Human Action*, published in 1864 (and reissued in 1965 by Harvard University Press).

In *Man and Nature*, Marsh spelled out his theory that it was degradation of the natural environment that led to the decline of earlier civilizations. Marsh had seen in his home state of Vermont and in countries around the world the effects of deforestation on soil, plant and animal life, and streams. He predicted that the Earth would become unfit for human habitation if people continued to destroy natural resources, which in turn would lead to impoverishment, crime, barbarism, and "perhaps even extinction of the species."

Marsh died in 1882 but is still remembered today as one of the first to write about the interrelatedness of all life, describing the concept of the web of life and what is now called the science of ecology. He made clear that nature by itself cannot necessarily restore or heal the destructive effects of human activities; people need to aid in the restoration process.

Further Reading: Curtis, Jane, Will Curtis, and Frank Lieberman. 1982. *The World of George Perkins Marsh.* Woodstock, VT: Countryman Press; Hanson, Susan, ed. 1997. *Ten Geographic Ideas that Changed the World.* New Brunswick, NJ: Rutgers University Press; Lowenthal, David. 1958. *George Perkins Marsh: Versatile Vermonter.* New York: Columbia University Press; George P. Marsh. 1965. Reprint. *Man and Nature; or, Physical Geography as Modified by Human Action.* Cambridge, MA: Belknap Press of Harvard University. Original edition: New York: C. Scribner, 1864; Meyer, William B. 1996. *Human Impact on the Earth.* Cambridge: Cambridge University Press.

Mendes, Chico (1944–1988)

Francisco "Chico" Alves Mendes Filho, known as Chico Mendes, was born in 1944 in the partially deforested ranchland of Porto Rico, Brazil. His family worked and lived on a rubber plantation, growing their own food or harvesting it from the nearby forest. Chico, the oldest of six children surviving (out of seventeen), learned early how to tap rubber trees and collect the latex. He also learned in early childhood how rubber barons and other wealthy landowners and military leaders in Brazil had long suppressed and exploited peasants, indigenous populations, and poor immigrants. By the time he was in his early twenties, he was trying to organize rubber tappers to demand schools, better working condition, and fair prices for the goods they had to buy from company stores. Mendes left the rubber plantation in 1971 and taught adults at a nearby government school.

About the same time, the Brazilian government was building roads through the rainforest to the state of Acre, where the rich rainforests are located, offering incentives to attract industry and cattle ranchers. Firsthand, Mendes saw hired workers burn and cut down forests, driving rubber tappers and poor farmers from the land. Mendes, with the help of activist Catholic priests, joined several other grassroots leaders and began to teach rubber tappers their rights and to organize unions—a dangerous undertaking since many leaders and members were killed or beaten by ranchers or their hired thugs.

By the early 1980s, Mendes was actively recruiting unionists and working with a national organization of rural workers, who held a conference in Brasilia in 1984. The conference brought together more than 4,000 delegates. During the conference, Mendes promoted a forest region plan that would eventually become the basis for *extractive reserves*—those regions set aside to preserve the forest. Although his plan was not adopted that year, Mendes continued his fight—in spite of threats on his life and the murder of other activists, including rural workers, priests, and lawyers.

In 1987, Mendes received the United Nation's Global 500 Award and the Better World Society medal for environmental conservation. His efforts also were featured in a major *New York Times* article and in other major newspapers and magazines in the United States. But Mendes was in constant danger from enemies in his own country who were enraged because of his success in preventing rainforest destruction, and in December 1988 he was murdered by cattle ranchers. His death brought international attention to the Amazon rainforest and the efforts of many brave people to save it.

Further Reading: Revkin, Andrew. 1990. *The Burning Season: The Murder of Chico Mendes and the Fight for the Amazon Rain Forest.* Boston: Houghton Mifflin; Shoumatoff, Alex. 1990. *The World Is Burning: Murder in the Rain Forest.* Boston: Little, Brown.

Montiel Flores, Rodolfo (1955–)

Rodolfo Montiel Flores is a peasant farmer in Mexico who with his colleagues organized Campesinos Ecologistas de la Sierra de Petatlán y Coyuca de Catalán (Farmer Ecologists from the Mountains of Petatlán and Coyuca de Catalán). Their purpose is to protest commercial logging in the mountainous region north of Acapulco.

In 1995, Boise Cascade, a U.S. company, began logging in the Sierra de Petatlán, contracting with the area's union of *ejidos*, agricultural communal groups that own most of Mexico's forest lands. Because of the North American Free Trade Agreement (NAFTA) and the Mexican Forest Reform Law of 1997, foreign investors are allowed to waive protections for biodiversity and water quality, which apparently was the case with Boise Cascade.

When the company pulled out in 1998, local logging operations took over, supported by leaders of the ejidos who have strong ties with authorities and have been accused of numerous corrupt practices and brutality.

The disturbance to the forest watershed has drastically decreased and polluted the region's water supply, affecting poor, agricultural villages the most and prompting Montiel and about 100 other farmers to register complaints with government officials. But those complaints have been ignored. As a result, the campesinos set up roadblocks to stop logging trucks, angering the loggers. Montiel and his family, whose lives were threatened, were forced to flee their village, but in 1999 Montiel and a friend were captured by the military, tortured, and eventually convicted on trumped-up charges of trafficking in weapons and narcotics.

In 2000, Montiel was awarded the Goldman Environmental Prize (see entry in this chapter) for courageous activism and was named an Amnesty International prisoner of conscience. He explained the reasons for his activism in a statement released by the Goldman Environmental Foundation:

> When there are trees on one ridge, and also on a neighboring ridge, the clouds knock together and there is rainfall, but when one ridge has no trees, the clouds just pass by and only one or two drops fall, and as a result crops are lost, hurting the campesinos and the professional people who eat the crops the campesinos produce. The government has a solution in hand, not persecuting the campesinos, or sending them to jail or killing them, but rather backing them up with productive projects that do not harm the environment, supporting first the reforestation of affected areas, and immediately requiring those responsible for the destruction to rebuild them, and providing support in the form of tractors, water pumps, chicken farms, pigs, and for the cattle raisers, feed grinders for grinding sugar cane, sorghum, and cornstalks, so that they can produce feed for their cattle, and not destroy the forests, converting them into dried up wasteland, as has been done. . . . [E]ach tree they cut down is like an exploding bomb and the springs disappear, the sea rises, and along with deforestation and burning, the flora dies, that is, they kill the ecosystem and our soils erode and each day they become less fertile and the campesino loses. . . . That is why I invite every-

body to share the water we have to drink and the food produced by the earth, let us look at it is if it were yours, not to destroy, but to build and to demand respect for our heritage. Let us become aware, because it's for the good of your children, your grandchildren, and all the generations; since we are only passing through, at least we can leave them some pure air to breathe. This is the respectful wish of your friend—Rodolfo Montiel Flores.

Further Reading: Dillon, Sam. 2000. "Jailed Mexican Wins Environmental Prize." *New York Times* (online version), April 5. http://www.goldmanprize.org/press/clippingItem.cfm?pressClipID=31. Accessed March 4, 2001; Montiel Flores, Rodolfo. "Statement" (translated from Spanish). April 5, 2000. In press release: "Ecological Situation in the Townships of Petatlán and Coyuca de Catalán, in the state of Guerrero, México." Goldman Prize website. http://www.goldmanprize.org/press/pressReleaseItem.cfm?prID=15. Accessed March 4, 2001; Ross, John. 2000. "Defending the Forest and Other Crimes." *Sierra* (online version). July-August. http://www.sierraclub.org/sierra/200007/Mexico.asp. Accessed March 4, 2001; Smith, James F. 2000. "Mexico's Forest at a Watershed." *Los Angeles Times,* May 16, A1.

Myers, Norman (1936–)

A leading British ecologist and an international consultant in environmental protection and development, Norman Myers focuses his research on resource relationships between the developed and developing countries. A winner of numerous environmental awards, Myers has been a consultant to various United Nations agencies, the World Bank, the U.S. National Academy of Sciences, the Smithsonian Institution, and many other organizations, and he has analyzed the problems of deforestation in sixty tropical countries.

Myers helped alert the world to the loss of biodiversity by originating the biodiversity strategy based on what he called "hot spots." Born on a sheep farm in northern Pennines, England, Myers attended Oxford University. Then in 1958, at the age of twenty-two, he went to Kenya as a colonial administrator. He described the job as "marvelous," saying the work was done in quite a democratic fashion.

I was posted to an area of African tribes, first of all some Kipsigaos and then the Masai tribe, which was fantastic.

And really I could only do anything with the active collaboration of the tribal elders and chiefs. I would have a certain amount of money from central government and I would say to them, "Do you want a cattle dipping station, or a new school, or a community hall, or a road, or whatever?" And for every one dollar that came from central government, they had to make one dollar themselves. And we'd sit under a thorn tree and debate this for endless days, and drink milk and blood, and so on. It was such fine work. And it also had a lot to do with wildlife (Institute of International Studies 1998).

While in Africa, Myers began taking photographs of wildlife and found he was able to sell them commercially, thus becoming a professional photographer. Many of his photograph collections have been published in book form. Myers has written numerous magazine articles and books covering ecology and rainforest issues, among them *The Primary Source: Tropical Forests and Our Future* (1992) and *Perverse Subsidies: Tax $s Undercutting Our Economics and Environments Alike* (1998). He is a fellow of Green College, Oxford University, and holds visiting professorships at the Universities of California, Texas, Michigan, Cape Town and Utrecht, as well as Stanford, Harvard, and Cornell, and he is a foreign member of the U.S. National Academy of Sciences.

Further Reading: Myers, Norman. 1999. "What We Must Do to Counter the Biotic Holocaust." *International Wildlife Magazine* (online version). March-April. http://www.nwf.org/intlwild/1998/contma99.html. Accessed March 4, 2001; Myers, Norman, ed. 1993. *The Gaia Atlas of Planet Management.* New York: Anchor Books/Doubleday; Institute of International Studies, University of California, Berkeley. November 11, 1998. "The Journey of an Environmental Scientist: Conversation with Norman Myers." Conversations with History website. http://globetrotter.berkeley.edu/people/Myers/myers-con0.html. Accessed February 2, 2001.

Nugkuag Ikanan, Evaristo (1950–)

A member of the Aguaruna Indian tribe, Evaristo Nugkuag is a grassroots organizer in Peru. He was educated at missionary schools and, like Chico Mendes, saw as he was growing up how logging, cattle ranching, and mining were destroying his homeland. He studied medicine in Lima, and in the late 1970s he began rallying the Aguaruna people to gain control of their land.

He soon realized that all Native people living in the Peruvian Amazon faced the same threats.

In 1981, Nugkuag brought together thirteen tribal groups representing more than half of the indigenous people of Peru's rainforest to find ways to protect their people's basic human rights and to save their traditional lands from the encroachment of timber and mining companies as well as from drug dealers. In 1984, he negotiated with tribal leaders in other nations to develop the Coordinating Organization for Indigenous Bodies in the Amazon Basin. The organization now represents more than half of the indigenous people in Amazonia, including tribal groups not only in Peru but also in Colombia, Brazil, Ecuador, and Bolivia. The organization works to defend indigenous rights, land, and resources and to bring about political representation and education for indigenous people.

As a grassroots leader, Nugkuag has received several international awards, one of which was the 1986 Right Livelihood Award for organizing the Native peoples of the Amazon basin to protect their rights. In 1991, he received the Goldman Environmental Prize (see entry in this chapter). During recent years, Nugkuag has been working on a human development strategy for the Aguaruna and Huambisa people. Among his initiatives is the development of Amazonian forest resources without logging.

Further Reading: "Evaristo Nugkuag." 1999. Goldman Environmental Prize website. http://www.goldmanprize.org/recipients/byyear.html.

Paiakan, Paulo (1954?–)

Paulo Paiakan, who has no record of his birth date but probably was born about 1954 or 1955, is a chief of a Kayapo tribe in the Amazon rainforest and has brought together indigenous people and environmental groups to work cooperatively in efforts to save the forests and the people who live there. Descended from a long line of chiefs, Paiakan knew early in his life that his special destiny was to "go out into the world to learn what was coming" to his people, as he told a *Parade Magazine* journalist (Whittemore 1992, pp. 4–7).

During his teenage years, Paiakan saw the destruction of forestland to construct the Trans-Amazonian Highway and additional devastation brought by the loggers, ranchers, and miners who followed. The chief tried to convince villagers to leave their homes and move farther into the forest, where they would be able to maintain their way of life. Most villagers thought the forest could not be destroyed, and only about 150 people agreed to move away, finally settling in Aukre, which is in the state of Pará, in northern Brazil.

Although the Aukre villagers found a better life, many other Kayapo were adversely affected by the pollution and disease that came with the invaders. Paiakan, who had attended a missionary school for his early education, decided to fight for his people and during the 1980s went to the state capital, Belém, where he learned to live like white people and to speak their language, Portuguese. His mission was to educate government leaders about threats to indigenous people and to convince his people to protect their forest lands and to ward off the temptation to trade their timber for flashy manufactured goods. Over the decade, Paiakan learned to operate a video camera so he could film the deforestation activities and show them to his people; he organized protests against a planned hydroelectric dam that would flood vast areas of rainforest; he traveled to the United States, Europe, and Japan to publicize the threats to Amazon Indians.

In the early 1990s, Paiakan began efforts to convince indigenous leaders that selling rainforest products was a way to preserve the forest and earn income. He helped them to organize and made arrangements for them to attend a conference of NGOs held at the same time as the Earth Summit in Rio de Janeiro, in June 1992. His work to maintain the Kayapo culture and at the same time convince the outside world to work with nature rather than to overcome it goes on.

Further Reading: Whittemore, Hank. 1992. "I Fight for Our Future." *Parade Magazine*, April 12: 4–7.

Pinchot, Gifford (1865–1946)

Born in Connecticut, Gifford Pinchot served as the first chief of the U.S. Forest Service. Pinchot was a member of a wealthy family of merchants, politicians, and land owners and was educated at some of the best eastern schools. His father urged him to make forestry his profession, but no U.S. university offered such a degree at that time. After graduating from Yale, Pinchot studied forestry in France, returning to the United States to work as a forester.

In 1896, Pinchot became part of the National Forest Commission, which was set up to identify forest areas for possible forest reserves (now known as national forests, some of which are rainforests). In 1898, Pinchot became head of the Division of Forestry in the Department of Interior. When President Theodore Roosevelt took office, the management of forest reserves was transferred from the Department of the Interior to the Department of Agriculture, and Roosevelt named as chief forester his friend Pinchot.

With Roosevelt's support, Pinchot restructured and professionalized the management of the national forests. Until the early 1900s, there had been no system in the country for managing private or government-owned timberland. As Pinchot observed, "to waste timber was a virtue, not a crime." Lumbermen "regarded forest devastation as normal and second growth as a delusion of fools." Generally considered the "father" of American conservation because of his unrelenting concern for the protection of the American forests, Pinchot emphasized long-term forest management and wise use of timber resources. Under Pinchot's guidance as forest chief from 1905 to 1910, the number of national forests more than doubled. Pinchot also founded the Society of American Foresters and the School of Forestry at Yale University and wrote numerous reports and books on conservation as well as an autobiography, which has been reprinted several times and explains his views of the U.S. conservation movement.

Further Reading: Pinchot, Gifford. 1988. Reprint. *Breaking New Ground.* 1998 edition, Washington, DC: Island Press. Original edition: New York: Harcourt Brace Jovanovich, 1947; "Pinchot, Gifford." 2000. Microsoft Encarta Online Encyclopedia 2000. http://encarta.msn.com

Raven, Peter H. (1936–)

Noted botanist and rainforest expert, Peter H. Raven is director of the Missouri Botanical Garden in St. Louis and is also Engelmann Professor of Botany at Washington University, St. Louis. *Time* magazine called him a "Hero for the Planet" in 1999. A leading authority on tropical forests and their inhabitants, Raven is known worldwide for his groundbreaking research. In his position at the Missouri Botanical Garden, he has employed over thirty botanists and other specialists who are assigned to work throughout the tropics, researching and cataloging new tropical plants. Specimens are collected and brought to the St. Louis herbarium for further study.

Born in China in 1936, Raven was educated at the University of California, Berkeley, and the University of California, Los Angeles. He holds honorary degrees from numerous universities and has received dozens of awards and honors from scientific, botanical, horticultural, and conservation organizations. He also has written hundreds of scientific papers on tropical plants and animals and has served on numerous editorial boards. Raven has coauthored or coedited numerous books and articles.

Today Raven focuses much of his attention on what he considers the menace of a "sixth extinction," a potential mass extinction of living organisms that could be brought about by the mushrooming human population and by human carelessness and commerce. In the new millennium, Dr. Raven is calling for an "age of biology," within which humans strive to fully understand the diversity of the Earth's living organisms and to use the properties of those organisms as a means to develop sustainability and conserve biodiversity.

In 2000, Raven became president of the American Association for the Advancement of Science, and President Bill Clinton awarded him a National Medal of Science. The award was established by Congress in 1959 and is administered by the National Science Foundation. It honors individuals for contributions to the present state of knowledge across a variety of science frontiers. Raven was recognized as one of the world's leading authorities on plant systematics and evolution and as a leader in international efforts to preserve biodiversity.

Further Reading: Palevitz, Barry, and Ricki Lewis. 1999. "A Conversation with Peter Raven." *The Scientist.* September 13: 1; Rosenblatt, Roger. April 19, 1999. "The World Is His Garden. Better Tread Carefully." Time.com. http://www.time.com/time/reports/environment/heroes/heroesgallery/0,2967,raven,00.html. Accessed March 5, 2001.

Red List of Threatened Species (1994–)

Since 1994, the International Union for the Conservation of Nature—also known as the World Conservation Union—has issued Red Lists of threatened species, which are considered the most comprehensive lists available. Over the years, the Red Lists have become larger and more complex. A total of 18,276 species and subspecies are included in the 2000 Red List, which notes that rainforest destruction, crop and livestock farming, timber plantations, mining, fisheries, logging, harvesting, and industrial and urban development have contributed to loss of habitat and species worldwide. According to the 2000 IUCN Red List of Threatened Species, critically endangered species increased dramatically from the last assessment, made in 1996. Indonesia, India, Brazil, and China are among the countries with the most threatened mammals and birds, while plant species are declining rapidly in South and Central America, Central and West Africa, and Southeast Asia. An analysis of the report, published in booklet form, called for widespread mobilization of human and financial resources to address this crisis. The 2000 IUCN Red List is available on CD-ROM and is searchable on its own website: http://www.redlist.org.

Roddick, Anita (1942–)

"The Indians Are the Custodians of the Rainforests. The Rainforests Are the Lungs of the World. If They Die, We All Die." These words are typical of the slogans printed on semi-trucks that transport products from The Body Shop International, a cosmetic manufacturer and retailer founded by Anita Roddick in Littlehampton, England. Roddick started the business in a storefront shop in 1976 with several goals in mind: to produce and sell "natural" skin and hair care products, to foster education, and to promote the concept of service to others. Rather than using the usual marketing techniques, the company has depended on word-of-mouth advertising, store window campaigns, community outreach programs, and mail order catalogs to reach its customers. Expanding rapidly, the company now includes more than 800 shops in forty countries around the world, including Canada and the United States.

To Roddick, principles are as important as profits, so she and her staff spend a lot of time on quality-of-life issues such as environmental concerns and human rights, actively supporting groups like Cultural Survival and various international environmental organizations. The Body Shop is as well known for its environmental activism as for its products. Roddick and her company also promote the concept of "trade-not-aid" as a positive approach to solving economic problems in the developing world. Rather than giving handouts, the company helps communities around the world obtain the resources they need to support themselves. Trade-not-aid projects have helped indigenous groups in the tropics develop and market sustainable forest products.

Further Reading: "The Body Shop: Truth & Consequences." 1995. *Drug & Cosmetic Industry* (February) 54–59; Chatzky, Jean Sherman. 1992. "Changing the World." *Forbes* 2 (March): 83–84; Elmer-Dewitt, Philip. 1993. "Anita the Agitator." *Time* 25 (January): 52–55; Roddick, Anita. 1999. "Trading with Principles." Address to the International Forum On Globalisation's Tech-In, Seattle, Washington. November 27. Gifts of Speech website. http://gos.sbc.edu/r/roddick.html. Accessed March 4, 2001.

Wilson, Edward O. (1929–)

A leading American entomologist, Edward O. Wilson is a Pellegrino University Professor at Harvard University and an expert in myrmecology, the study of ants and their social systems. Twice he has won the Pulitzer Prize, for his books *On Human Nature* (1978) and *The Ants* (1990). He is known worldwide for his controversial work *Sociobiology: The New Synthesis* (1975), which contends that social behavior has a biological basis. Over the years, social scientists and other scholars have bitterly condemned Wilson's claim, but in the introduction to the twenty-fifth anniversary edition of the book, Wilson points out that research in human genetics and neuroscience has strengthened the case for a biological understanding of human nature. In fact, human sociobiology, now frequently called evolutionary psychology, is a field of study based on biology and social science data.

Born in Birmingham, Alabama, Wilson spent much of his childhood in the rural areas of northern Florida and Alabama, where he not only roamed the woods and enjoyed hunting and fishing, but also, in his words, "cherished natural history for its own sake and decided very early to become a biologist." Wilson was educated at the University of Alabama and Harvard University and was appointed to the Harvard faculty in 1956. He has conducted research in rainforests around the world and has received numerous awards from scientific organizations in the United States, Sweden, France, and other countries.

Wilson is a member of the Club of the Earth, which is made up of nine prominent American scientists who believe that species extinction of unimaginable magnitude will occur if broad conservation efforts are not undertaken in the near future. As he points out in an article for *Defenders Magazine:*

> Before the coming of humanity, each species and its descendants lived for somewhere between 1 and 10 million

years. Each species can be viewed as an encyclopedia of genetic information. The typical higher organism, anything from an amoeba on up to a human being, contains from 1 to 10 billion nucleotide pairs or genetic letters. For a mouse, that is roughly equivalent to all the information in all editions of the *Encyclopaedia Britannica* published since 1768. These occur in unique combinations. They prescribe unique traits in anatomy and behavior and biochemistry. And those in turn exquisitely adapt the species to the ecosystem in which it lives. When a species goes extinct, all that heritage is lost (Wilson 1993).

In spite of the increasing threat to biodiversity, often due to rainforest loss, Wilson is optimistic that people of varied persuasions can come together in their efforts to sustain the earth for future generations.

Further Reading: Branfman, Fred. 2000. "Living in Shimmering Disequilibrium." *Salon Magazine,* (online version) April 22. http://www.salon.com/people/feature/2000/eowilson/. Accessed March 4, 2001; Wilson, Edward O., ed. 1988. *Biodiversity.* Washington, DC: National Academy Press; Wilson, Edward O., ed. 1993. "The Threatened Biosphere." *Defenders* (online version). Summer. http://www.defenders.org/biobi01.html. Accessed March 4, 2001.

Weyerhaeuser, Frederick (1834–1914)

Frederick Weyerhaeuser, who became known as "the Timber King," was born in Niedersaulheim, Germany. In 1852 he emigrated to the United States, working as a laborer in Erie, Pennsylvania. After marrying Elizabeth Bladel, he moved to Rock Island, Illinois, where he found work in a sawmill. Weyerhaeuser was able to save enough money to buy the sawmill and a timber yard in Rock Island in 1857. The business was a great success and he bought additional sawmills on the shores of the Mississippi.

In 1891, Weyerhaeuser moved to St. Paul, Minnesota, where his neighbor was James Hill, railroad tycoon, who for a low price had purchased millions of acres of forests from the U.S. government. Hill's interest was railroad construction, and knowing little about the timber industry, he sold more than 3 million acres to Weyerhaeuser. By the end of the nineteenth century, Weyerhaeuser owned more timberland than any other American. His company, Weyerhaeuser Company, is now a multinational corporation under the leadership of his grandson Philip, and it has come under frequent attack because of rainforest clear-cutting.

Further Reading: Ramsey, Bruce. 1999. "Philip Weyerhaeuser—Timberman with a Vision." *Seattle Post-Intelligencer Reporter,* December 29; Wandel, Joseph. 1979. *The German Dimension of Americana History.* Chicago: Nelson-Hall; "Weyerhaeuser, Frederick (1834–1914)." German Corner website. http://www.germanheritage.com/biographies/mtoz/weyerthaeuser.html. Accessed February 6, 2001; Yaron, Gil. 2000. "The Corporatization of B.C.'s Forests." *Canadian Dimension,* February: 1.

"Wise Use" Movement

The "wise use" movement got under way in the 1980s, but it is part of a larger and older property rights movement supported by numerous Americans who fear that the government is taking over excessive amounts of private land for public use or for species and environmental protection. Under the power of eminent domain, the U.S. government can take private property for a public purpose, but the Fifth Amendment to the U.S. Constitution states that private property shall not "be taken for public use, without just compensation." This "takings clause," as it is called, ensures that property owners are paid a fair price when government takes land for a public purpose. Historians say the clause also was meant to protect the landholdings of the wealthy who did not want property redistributed to the masses.

In 1988, a former environmental activist, Ron Arnold, and Alan Gottlieb, a political fund-raiser, proposed the "wise use" of natural resources, adopting a popular conservation motto established by Gifford Pinchot, the first chief of the U.S. Forest Service. In 1907, Pinchot had proclaimed that forests and other natural resources should not be exploited but conserved so that they would be available for future generations. However, today the wise use movement often opposes conservation efforts and especially denounces the Endangered Species Act.

Wise-use supporters advocate unlimited mining, logging, and grazing on public lands, insisting that these industries do not interfere with the preservation of natural resources. Some miners, loggers, cattle ranchers, and oil and gas producers want to open up more government-controlled acreage than is already available for private operations. Governors of most western states where these operations take place believe they, not the federal government, should take charge of managing the public lands within their boundaries.

Further Reading: Gay, Kathlyn. 1994. *Pollution and the Powerless.* Danbury, CT: Franklin Watts/Grolier; Gay, Kathlyn. 1996. *Saving the Environment, Debating the Costs.* Danbury, CT: Franklin Watts/Grolier; Libby, Ronald T. 1998. *Eco-Wars: Political Campaigns and Social Movements.* New York: Columbia University Press.

5

Data and Documents

In general, this chapter supplements and illustrates information provided in Chapters 1 and 2. It begins with a graph showing forest cover changes in various parts of the world from 1990 to 1995. Comparisons of tropical and temperate rainforest factors are shown in Table 1.

A section on temperate rainforests includes tables explaining how timber is harvested, various methods for saving timber by reducing wood and paper use, the sizes of Olympic trees in North America, and how life is sustained from a dead log.

Tropical rainforests are the topic of the next section, which includes tables showing common products from rainforest resources, a list of historical facts about chocolate production from rainforest cacao beans, graphs showing production and consumption of forest products worldwide, the twenty-five biodiversity hot spots, and a table showing the costs of dam projects financed with the help of World Bank loans.

The final section includes seven documents relating to tropical issues, including four examples of declarations and manifestos that have been widely disseminated on the Internet. Among these, a "side agreement" to the North American Free Trade Agreement, called the North American Agreement on Environmental Cooperation, deals primarily with environmental protections across national borders. Other documents apply to the preservation of biodiversity and endangered species. Finally, two rainforest-related documents from the UN Conference on Environment and Development are included: The Rio Declaration and a "Statement of Principles for a Global Consensus on the Management, Conservation and Sustainable Development of All Types of Forests."

Forest Cover

This graph showing forest changes is based on information from the World Resources Institute (WRI) and data compiled by the Food and Agriculture Organization of the United Nations. Forest cover as WRI explained, "consists of all forest area for temperate developed countries, and the sum of natural forest and plantation area categories for tropical and temperate developing countries" (WRI).

Figure 1. Forest Cover Changes from 1990 to 1995
(in 1,000 hectares)

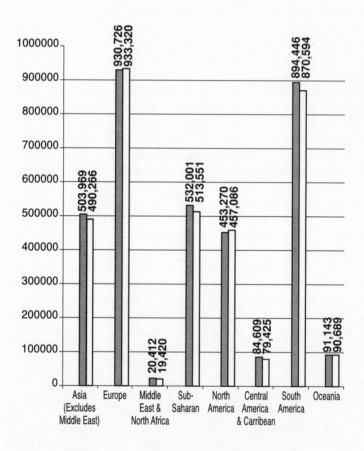

Sources: World Resources Institute, Food and Agriculture Organization of the United Nations

Table 1
Comparing Temperate and Tropical Rainforests

	TEMPERATE RAINFORESTS	TROPICAL RAINFORESTS
LOCATIONS	Pacific Northwest	Intermittently from Brazil to Mexico
	British Columbia	Congo Basin
	Southeast Alaska	Parts of S.E. Asia
	Southern Australia	Parts of N.E. Australia
	New Zealand	Some Pacific Islands
	Chile	
	Small portions coastal Britain,	
	Norway, Japan	
RAINFALL	100 inches or more per year	100 to 400 inches per year
CLIMATE	Cool	Warm
BIOMASS	500 tons living biomass per acre	300 tons living biomass per acre
BIODIVERSITY	Low diversity	High diversity

Temperate Rainforests

Since the 1980s, many North Americans have become more aware of the rainforests on their own continent. Yet much remains to be learned about this dwindling resource. According to the estimates of some ecologists, only 40–50 percent of the original forests in North America remain.

Harvesting in U.S. Forests

In the United States, loggers use a variety of methods to harvest timber (see Table 2). Usually a silviculturalist, or timber manager—someone who manages a forest so that it will continue to yield timber, determines the harvesting system. A timber manager determines a harvest method by the type of ecosystem and the tree species in that system. A clear-cut method might be used, for example, if plenty of light is needed for natural regrowth of trees that do not reproduce well in heavy shade. Clear-cutting also may be less costly than other harvesting systems, because after a clear-cut, seedlings planted by hand are apt to mature faster (and bring a quicker profit) than those allowed to grow back in a natural process. On the other hand, clear-cutting is not always the least expensive harvesting method in the long run, especially if roads have to be constructed into logging sites. Frequently, logging companies in the Pacific Northwest clear-cut large areas of forest, but this can have a major impact on the exposed stands of forest on either side. Many controversies have erupted over excessive logging in the temperate rainforests, and

Table 2
Harvesting and Logging Systems

Harvesting

Clearcutting. Loggers cut all trees in an area ranging from 5 acres to several thousand acres; the U.S. Forest Service usually clearcuts between 40 and 60 acres.

Shelterwood cutting. Loggers cut one-half to two-thirds of the trees in a stand, leaving the rest of the trees to shelter seedlings.

Seed tree cutting. Loggers cut nearly all trees on each acre, leaving just a few as a source of seeds but not enough to shelter seedlings.

Selection cutting. Loggers select a group of trees or individual trees in a stand for cutting.

Logging

Tractor logging. Large tractors drag logs down slopes and along trails to logging trucks that transport logs to mills. This is usually the most economical logging method but it often causes soil erosion.

High-lead logging. Lead cables are used to haul logs to a truck. Although more expensive than tractor logging, this method does less damage to soil.

Skyline logging. Heavy cables stretched between two points carry suspended logs to a landing area, preventing soil damage but increasing logging costs.

Helicopter and balloon logging. Logs suspended from a helicopter or balloon are airlifted to transport sites—the most expensive type of logging.

Source: "The Art of Silviculture," The Citizens' Guide to Timber Management in the National Forests. Eugene, OR: Cascade Holistic Economic Consultants, May 1989, pp. 12-21.

a variety of factors, such as bans on logging to protect endangered species and less demand for timber, have led to decreases in timber harvests.

Reducing Wood and Paper Use

Whenever trees are cut, the timber is sawed, peeled, chipped, or burned to produce a variety of products. To reduce the use of wood from nonsustainable forests in North America, such as the ancient rainforests in British Columbia, hundreds of businesses have signed a WoodWise Business Pledge. With this pledge, these businesses indicate their commitment to avoiding the use of old-growth wood products and building markets for more sustainable alternatives. Their names are posted on the Internet at http://www.coopamerica.org/woodwise/directory.pdf. Other businesses and organizations are members of the Certified Forest Products Council, which is engaged in a mission to "conserve, protect and restore the world's forests by promoting responsible forest products buying practices throughout North America." According to its mission statement, the Council "promotes and facilitates the increased purchase, use and sale of third-party independently certified forest products. The Council also promotes the transition away from forest products originating in forests that have been identified as endangered through a scientifically credible, land-based assessment process. In addition, the Council encourages its members to promote the appropriate and efficient use of wood and wood fiber, and to support the development and use of alternative products." Consumers can learn about the Certified Forest Products Council and certified forest products worldwide at its website, http://www.certifiedwood.org/.

Paper is one of the most common products requiring wood, and paper recycling, which has been a popular activity for decades, has helped reduce wood consumption. During the 1990s, paper recycling increased steadily. In 1999 two million tons of paper fiber were recovered and recycled in the United States—double the amount recovered in 1989—according to the American Forest and Paper Association. Recycled fiber is used to make such products as cellulose insulation, molded pulp packaging, paper mulch, compost, and kitty litter. In addition to recycling, numerous groups encourage consumers to reduce their use of paper with a variety of suggestions (see Table 3).

Valuable Ancient Trees

A great variety of plant life thrives in temperate rainforests, as do diverse species of mammals and birds. (For detailed descriptions of plant and animal life found in ancient forests, see many of the books on temperate rainforests annotated in chapter 7.) For example, among the trees growing in the ancient forests of the Pacific Northwest are many types of evergreens, such as the Douglas fir, Sitka spruce, and western hemlock (Washington's state tree), and broadleaf trees such as the Red adler, sometimes mistaken for birch, and the Bigleaf Maple. The Olympic National Forest—a temperate rainforest—boasts some record size trees that have been recognized by the American Forestry Association as the largest living specimens of the species in their list of approximately 750 national champions.

One of the most valuable trees in an ancient forest is the Douglas fir. After hundreds of years of life, a Douglas fir tree may die of old age and topple over. But the log that rests on the forest floor or drops into a stream continues to provide life, acting as a host for countless living creatures and serving many other functions (see Table 5). If the ecosystems that support these trees are destroyed, Americans stand to lose a great deal.

Table 3
Suggestions to Reduce Paper Use

At home:
- Use durable dishes, mugs, glasses, and flatware rather than disposable items.
- Use cloth towels to dry hands and dishes.
- Use washable cloth napkins, placemats, and table coverings instead of disposable paper ones.
- Use dispensers for sugar, sweeteners, and other condiments instead of individual packets of them.
- Buy and use recycled paper products such as tissues, toilet paper, cardboard packaging, and printer paper.
- Use the back of envelopes for taking telephone messages or making shopping lists.
- Use old newspapers for wrapping items to be shipped or mailed.
- Take your own shopping bags to carry groceries from markets.
- When possible buy used furniture or new wood items that have the Forest Stewardship Council label or other certification that the wood is from managed forests.

In the office:
- Save paper that is blank on one side for reuse with printers and copy machines.
- Make double-sided photocopies.
- Print only the number of copies necessary.
- Limit the distribution of "hard-copy" correspondence and reports to only those who really need them.
- Route one hard copy to several readers.
- Use electronic mail or bulletin boards for sending and receiving information.
- Eliminate unnecessary layers of packaging.
- Shred or crumple waste paper for use as packing material.
- Reuse boxes and packaging for outgoing shipments.
- Store information on computer discs whenever possible.

Table 4
Record Olympic Trees

TREES	CIRCUMFERENCE	HEIGHT	SPREAD	LOCATION
Alaska Cedar	452 inches	120 feet	27 feet	Along the Hoh River Trail near Cougar Creek
Douglas Fir	533.5 inches	212 feet	47.5 feet	Queets River Trail to Klochman Rock Trail at Coal Creek
Douglas Fir	448 inches	298 feet	64 feet	South Fork Hoh River Trail
Grand Fir	229 inches	251 feet	43 fee	On the Duckabush River Trail
Subalpine Fir	253 inches	129 feet	22 feet	Near Cream Lake, head of Hoh River drainnage
Western Hemlock	270 inches	241 feet	67 feet	Along the Hoh River Trail near Cougar Creek
Western Hemlock	316 inches	202 feet	47 feet	On Wynoochee Trail
Vine Maple	35 inches	62 feet	31 feet	On South Fork Hoh River Trail
Shore Pine	157 inches	43 feet	44 feet	By Deer Ridge Trail near Deer Park Campground

Table 5
Life From a Dead Log

On wet ground, a fallen log holds moisture and prevents soil erosion.
Plants and tree seedlings take root in moist, fertile soil near a fallen log or within the log itself.
Spiders use cracks in a log for nests.
Beetles bore into a log, making a way for other organisms to spread through the rest of the wood.
Termites build a maze of tunnels and chambers in the log.
Various scavengers feed on loose bark.
In a stream, a fallen log creates protected pools where fish spawn and rapids where insects thrive and serve as food for fish.
As a log rots, it becomes spongelike, holding moisture and helping to provide nutrients for soil and plants.

Tropical Rainforest Products

One of the most compelling reasons for preserving tropical rainforests is the great many benefits they provide. Along with wood products, rainforests supply medicinal plants and a variety of products, such as oils and resins, that are now used in industry or will be developed for industrial purposes in the future. Agricultural products from the rainforest are also invaluable. Some fruits, vegetables, and fibers that originated in rainforests have through crossbreeding become part of agricultural production in industrialized nations. Yet crop breeders still need a steady supply of genes from wild plants and primitive crops to develop hardier modern varieties—those that can resist disease or other threats.

Hundreds of other products that people use in their daily lives also come from tropical rainforests (see Table 6). Among the most familiar are cocoa and chocolate, made from the rainforest cacao tree, often called a forest "food of the gods" (see Table 7). The rainforest also produces paper, paperboard, and wood-based panels (see Figures 2 and 3) as well as fuelwood, charcoal, and roundwood (log) products (see Figure 6). The major producers and consumers of tropical forest–products are organized into the International Tropical Timber Organization (ITTO), which is described in chapter 6 (see Figures 4 and 5). Figure 6 compares fuel wood and charcoal consumption and roundwood, or log, consumption in non-industrialized and industrialized nations.

"Hot Spots"

Although tropical rainforests cover only a small portion of the world's land area, many of them are among the world's richest biological storehouses. But they face threats from logging (both illegal and sanctioned), dams, mining, cattle grazing, and increasing

Table 6
Examples of Tropical Rainforest Products

Numerous commercial Internet Web sites offer rainforest products for sale. A variety of rainforest organizations also promote rainforest products. This list is just a sampling of many common items that originate in rainforests.

Fibers, Gums, and Resins (and their uses)
- Bamboo (furniture)
- Copal (resin used in varnish)
- Jute (rope)
- Kapok (insulation)
- Raffia (baskets)
- Ramie (fabric)
- Rattan (furniture)
- Rubber latex (rubber products)
- Tung oil (wood finisher)

Foods and Beverages
- Avocado
- Banana
- Brazil nuts
- Breadfruit
- Cashew nuts
- Chicle gum
- Coconut
- Coffee
- Grapefruit
- Guarana
- Guava
- Mango
- Okra
- Papaya
- Peanuts
- Peppers
- Pineapple
- Tea

Houseplants
- Fiddle-leaf fig
- Mother-in-law's tongue
- Orchid
- Parlor ivy
- Philodendron
- Rubber tree plant
- Swiss cheese plant
- Zebra plant

Spices
- Allspice
- Black Pepper
- Cardamom
- Cayenne
- Chili
- Cinnamon
- Cloves
- Ginger
- Mace
- Nutmeg
- Paprika
- Sesame seeds
- Vanilla

Oils for Cosmetics, Perfumes, Flavorings, and Cough Drops
- Camphor
- Cascarilla
- Coconut
- Eucalyptus
- Palm
- Rosewood

Pharmaceuticals (and some of their uses)
- Curare (muscle relaxant for surgery)
- Diosgenin (birth control, steroids, arthritis treatment)
- Quassia (insecticide)
- Quinine (anti-malarial, pneumonia treatment)
- Reserpie (sedative, tranquilizer)
- Strophanthus (heart disease)
- Strychnine (emetic, stimulant)

Sources: "Our Tropical Connection," *Environmental Almanac*, Boston: Houghton Mifflin, p. 284; Rainforest Action Network undated brochure; "Tropical Forest Product List." 1994. Smithsonian Institution, http://www.si.edu/sites/educate/troprain/products.htm

Table 7
Food of the Gods

Chocolate, a product of the rainforest cacao plant, has been called the "food of the gods"—it is one of the world's most desired foods. The use of cacao to make a bitter tasting beverage originated in the Amazon thousands of years ago and went through many transformations after European explorers took cacao beans back to their own lands. A few tidbits about cacao and chocolate show the worldwide popularity of this rainforest resource:

- During the twelfth century in Europe a frothy chocolate beverage was used in royal rituals such as marriage ceremonies.
- From a golden cup, the Aztec emperor Moctezuma drank xocatl, a bitter spicy beverage made from the cacao bean.
- Toasting to good fortune with a chocolate drink is a long standing Mexican tradition that has carried over to the United States.
- The Spanish are credited with initiating a sweetened chocolate drink, mixing cane sugar with processed cacao beans and water and creating a concoction thought to be an elixir or even an aphrodisiac.
- About the middle of the 1800s, cocoa butter was extracted from ground cocoa beans and the fat was added to ground bean paste to make a solid product for eating.
- In 1875, the Swiss developed solid milk chocolate.
- Early in the 1800s, Milton S. Hershey established a chocolate factory in Pennsylvania and built a company town known as Hershey.
- Chocolate bars and other types of chocolate candies are widely popular in North America and chocolates are a favorite gift during holidays, birthdays, and other special occasions.
- Chocolate festivals are common across the United States—an example is the annual festival held in Burlington, Wisconsin, where a Nestlé Food Company factory is located.
- The Empress Hotel in Victoria, British Columbia, Canada, celebrates chocolate with its Death by Chocolate Buffet served every evening in the hotel's lobby lounge.

Source: Kathlyn Gay and Martin K. Gay. 1996. Encyclopedia of North American Eating & Drinking Traditions, Customs & Rituals. Santa Barbara, CA: ABC-CLIO.

population, and many forests include "hot spots," a term coined by British ecologist Norman Myers in 1988 to identify places where large numbers of unique species, found nowhere else on Earth, are concentrated in small areas. Assuming that not all of the world's biologically diverse areas can be saved, Myers's strategy concentrates on protecting the hot spots—the world's unique terrestrial and plant and animal species that inhabit 1.4 percent of the Earth's land area.

Today twenty-five locations are identified by Conservation International as hot spots. Special efforts are under way to preserve both hot spots and large areas of rainforest. In 2000, scientists representing Conservation International and other groups from around the world held a five-day conference entitled "Defying Nature's End: A Practical Agenda for Saving the Planet" to tackle the enormous problems facing the world's ecosystems. One of the strategies proposed was to build a Center for Biodiversity Conservation in each of the prime hot spots.

Figure 2. **Average Annual Forest Products Production 1966–1998 Paper and Paperboard (in metric tonnes)**

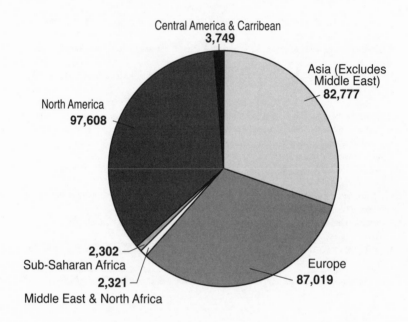

Central America & Carribean
3,749

Asia (Excludes Middle East)
82,777

North America
97,608

2,302
Sub-Saharan Africa
2,321
Middle East & North Africa

Europe
87,019

Sources: World Resources Institute, Food and Agriculture Organization of the United Nations

Figure 3.   **Average Annual Forest Production**
Wood Based Panels 1996–1998 (in 1,000 m3)

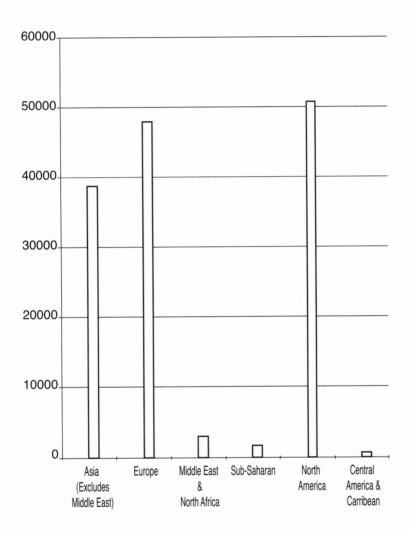

Source: World Resources Institute, Food and Agriculture
Organization of the United Nations, World Bank

**Figure 4.   Major Tropical Log Producers ☐
among ITTO Members (in 1,000 m3)**

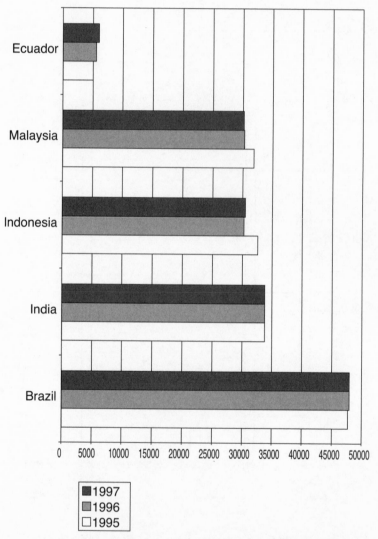

Source: International Tropical Timber Organization

Figure 5. Major Tropical Log Consumers among ITTO Members (in 1,000 m3)

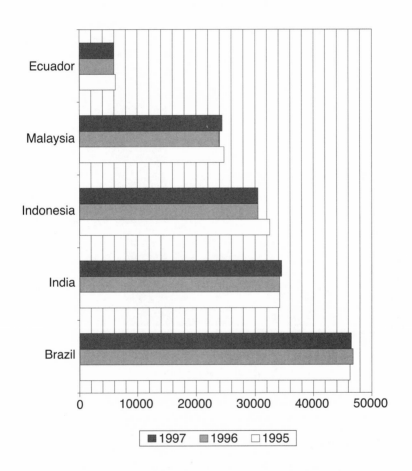

Source: International Tropical Timber Organization

Figure 6.   Wood Products Consumption 1996
Total Fuelwood & Charcoal 1,860 m3: Total Roundwood 1,498 m3

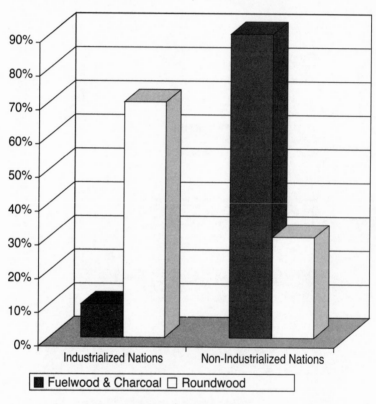

Fuelwood & Charcoal ☐ Roundwood

Source: State of the World's Forests 1999

World Bank Lending and the World Commission on Dams

The mission of the World Bank is to alleviate poverty in Third World nations by providing financial resources, knowledge, and partnerships in the public and private sectors, allowing people to help themselves. The World Bank provides nearly $16 million in loans annually to its client countries—funding that makes possible such development projects as dams and road building.

Funding for dam construction is one of the most controversial of the World Bank's lending activities (see Table 9). For several decades, environmental groups and indigenous people in rainforest areas have been criticizing the World Bank for its loans

Table 8
25 BIODIVERSITY HOTSPOTS
(identified by Conservation International)

1. Tropical Andes
2. Mediterranean Basin
3. Madagascar/Indian Ocean Islands
4. Mesoamerica
5. Caribbean Islands
6. Indo-Burma
7. Atlantic Forest of Brazil
8. Philippines
9. Cape Floristic Region of South Africa
10. Mountains of South Central China
11. Sundaland
12. Brazilian Cerrado
13. Southwest Australia
14. Polynesia and Micronesia
15. New Caledonia
16. Choco/Darien/Western Ecuador
17. Western Ghats & Sri Lanka
18. California Floristic Province
19. Succulent Karoo
20. New Zealand
21. Central Chile
22. Guinean Forests of West Africa
23. Caucasus
24. Eastern Arc Mountains, Coastal Forests of Kenya and Tanzania
25. Wallacea

Source: Conservation International. "Where Are the Hotspots." May 19, 2000 (Reprinted with permission). http://www.conservation.org/Hotspots/where.htm

for projects, such as large dams, that are supposed to bring economic development to poor nations but often prove to be harmful to local people and ecosystems.

This opposition seems to have made an impact. From 1970 to 1985, the World Bank provided loans for an estimated 3 percent of the world's new dam projects. But from 1985 to 1995 the World Bank's rate of involvement fell to about 2 percent, and by 2000 the bank's funding involved only about 1 percent of new dam projects worldwide. In recent years, World Bank spending has been focused primarily on dam rehabilitation and safety. A November 2000 report issued by the World Commission on Dams and supported by the World Bank found that although dam projects have brought some benefits, they have also been detrimental to people and the environment. This harm is described in the "Dam Building" section in Chapter 2 and in several of the documents presented below.

Table 9
World Bank Spending on New Dam Projects
1993–1999 (in million U.S. Dollars)

Year	Projects	Total Cost*	Total Loan	Dam Cost	Loan for Dam
1993	3	1,371.70	622.9	308	138.4
1994	4	3,827.90	1438	737.6	268.4
1995	6	4,610.80	1343	822.8	227
1996	2	501.6	270	81.6	43.8
1997	3	952.2	361	99	62.7
1998	2	1,404.80	203	292.4	30.7
1999	3	1,395.90	591.2	140.1	54.9
Total:	23	14,064.9	4,829.1	2,4815.5	825.9

*Total cost includes the amount spent for the dam itself, power equipment, irrigation canals, and any other construction activities.
Source: Statistics on the World Bank's Dam Portfolio. 2000. http://www.worldbank.org

Documents on Rainforests

A variety of documents and resolutions have been published since the 1980s to focus public attention on diverse environmental concerns, among them rainforest issues. A few of the most significant are reprinted here. Some are legal documents; others are public manifestos that have no legal standing but help illuminate the many problems associated with the destruction of rainforests, such as logging in Amazonia or the problems indigenous people have faced due to projects financed by the World Bank.

World Scientists' Warning to Humanity

"World Scientists' Warning to Humanity" is a document of the Union of Concerned Scientists, an organization of scientists and citizens seeking solutions to environmental problems including forest devastation, biodiversity loss, and climate change. In November 1992, 1,700 of the world's leading scientists, including the majority of Nobel laureates in the sciences, issued an appeal for fundamental changes to protect the world's critical resources. The document was written by the late Henry Kendall, former chair of the union's board of directors.

Introduction
Human beings and the natural world are on a collision course. Human activities inflict harsh and often irreversible damage on the environment and on critical resources. If not checked, many of our current practices put at serious risk the future that we wish for human society and the plant and animal kingdoms,

and may so alter the living world that it will be unable to sustain life in the manner that we know. Fundamental changes are urgent if we are to avoid the collision our present course will bring about.

The Environment

The environment is suffering critical stress:

The Atmosphere. Stratospheric ozone depletion threatens us with enhanced ultraviolet radiation at the earth's surface, which can be damaging or lethal to many life forms. Air pollution near ground level, and acid precipitation, are already causing widespread injury to humans, forests, and crops.

Water Resources. Heedless exploitation of depletable ground water supplies endangers food production and other essential human systems. Heavy demands on the world's surface waters have resulted in serious shortages in some 80 countries, containing 40 percent of the world's population. Pollution of rivers, lakes, and ground water further limits the supply.

Oceans. Destructive pressure on the oceans is severe, particularly in the coastal regions which produce most of the world's food fish. The total marine catch is now at or above the estimated maximum sustainable yield. Some fisheries have already shown signs of collapse. Rivers carrying heavy burdens of eroded soil into the seas also carry industrial, municipal, agricultural, and livestock waste—some of it toxic.

Soil. Loss of soil productivity, which is causing extensive land abandonment, is a widespread by-product of current practices in agriculture and animal husbandry. Since 1945, 11 percent of the earth's vegetated surface has been degraded—an area larger than India and China combined—and per capita food production in many parts of the world is decreasing.

Forests. Tropical rain forests, as well as tropical and temperate dry forests, are being destroyed rapidly. At present rates, some critical forest types will be gone in a few years, and most of the tropical rain forest will be gone before the end of the next century. With them will go large numbers of plant and animal species.

Living Species. The irreversible loss of species, which by 2100 may reach one-third of all species now living, is especially serious. We are losing the potential they hold for providing medicinal and other benefits, and the contribution that genetic diversity of life forms gives to the robustness of the world's biological systems and to the astonishing beauty of the earth itself. Much of this damage is irreversible on a scale of centuries, or permanent. Other processes appear to pose additional threats. Increasing levels of gases in the atmosphere from human activities, including carbon dioxide released from fossil fuel burning and from deforestation, may alter climate on a global scale. Predictions of global warming are still uncertain—with projected effects ranging from tolerable to very severe—but the potential risks are very great.

Our massive tampering with the world's interdependent web of life—coupled with the environmental damage inflicted by deforestation, species loss, and climate change—could trigger widespread adverse effects, including unpredictable collapses of critical biological systems whose interactions and dynamics we only imperfectly understand.

Uncertainty over the extent of these effects cannot excuse complacency or delay in facing the threats.

Population

The earth is finite. Its ability to absorb wastes and destructive effluent is finite. Its ability to provide food and energy is finite. Its ability to provide for growing numbers of people is finite. And we are fast approaching many of the earth's limits. Current economic practices which damage the environment, in both developed and underdeveloped nations, cannot be continued without the risk that vital global systems will be damaged beyond repair.

Pressures resulting from unrestrained population growth put demands on the natural world that can overwhelm any efforts to achieve a sustainable future. If we are to halt the destruction of our environment, we must accept limits to that growth. A World Bank estimate indicates that world population will not stabilize at less than 12.4 billion, while the United Nations concludes that the eventual total could reach 14 billion, a near tripling of today's 5.4 billion. But, even at this moment, one person in five lives in absolute poverty without enough to eat, and one in ten suffers serious malnutrition.

No more than one or a few decades remain before the chance to avert the threats we now confront will be lost and the prospects for humanity immeasurably diminished.

Warning

We the undersigned, senior members of the world's scientific community, hereby warn all humanity of what lies ahead. A great change in our stewardship of the earth and the life on it is required, if vast human misery is to be avoided and our global home on this planet is not to be irretrievably mutilated.

What We Must Do

Five inextricably linked areas must be addressed simultaneously:

We must bring environmentally damaging activities under control to restore and protect the integrity of the earth's systems we depend on.

We must, for example, move away from fossil fuels to more benign, inexhaustible energy sources to cut greenhouse gas emissions and the pollution of our air and water. Priority must be given to the development of energy sources matched to Third World needs—small-scale and relatively easy to implement.

We must halt deforestation, injury to and loss of agricultural land, and the loss of terrestrial and marine plant and animal species.

We must manage resources crucial to human welfare more effectively.

We must give high priority to efficient use of energy, water, and other materials, including expansion of conservation and recycling.

We must stabilize population.

This will be possible only if all nations recognize that it requires improved social and economic conditions, and the adoption of effective, voluntary family planning.

We must reduce and eventually eliminate poverty.

We must ensure sexual equality, and guarantee women control over their own reproductive decisions.

Developed Nations Must Act Now

The developed nations are the largest polluters in the world today. They must greatly reduce their overconsumption, if we are to reduce pressures on resources and the global environment. The developed nations have the obligation to provide aid and support to developing nations, because only the developed nations have the financial resources and the technical skills for these tasks.

Acting on this recognition is not altruism, but enlightened self-interest: whether industrialized or not, we all have but one lifeboat. No nation can escape from injury when global biological systems are damaged. No nation can escape from conflicts over increasingly scarce resources. In addition, environmental and economic instabilities will cause mass migrations with incalculable consequences for developed and undeveloped nations alike.

Developing nations must realize that environmental damage is one of the gravest threats they face, and that attempts to blunt it will be overwhelmed if their populations go unchecked. The greatest peril is to become trapped in spirals of environmental decline, poverty, and unrest, leading to social, economic, and environmental collapse.

Success in this global endeavor will require a great reduction in violence and war. Resources now devoted to the preparation and conduct of war—amounting to over $1 trillion annually—will be badly needed in the new tasks and should be diverted to the new challenges.

A new ethic is required—a new attitude towards discharging our responsibility for caring for ourselves and for the earth. We must recognize the earth's limited capacity to provide for us. We must recognize its fragility. We must no longer allow it to be ravaged. This ethic must motivate a great movement, convincing reluctant leaders and reluctant governments and reluctant peoples themselves to effect the needed changes.

The scientists issuing this warning hope that our message will reach and affect people everywhere. We need the help of many.

We require the help of the world community of scientists— natural, social, economic, and political.

We require the help of the world's business and industrial leaders.

We require the help of the world's religious leaders.

We require the help of the world's peoples.

We call on all to join us in this task.

Source: Union of Concerned Scientists http://www.ucsusa.org/resources/warning.html
(Reprinted with permission)

Manifesto to the Population. Predatory Logging Threatens Amazonia (1992)

The Coalition Against Predatory Logging in the Amazon, which includes agricultural, environmental, educational, human rights groups and other nongovernmental organizations (NGOs), met in Brazil in 1992 and released a manifesto decrying centuries of "predatory exploitation of natural resources" in the Amazon. The declaration denounces the devastation of forests and the continued depletion of mahogany, which is known as "green gold" because it is the most valuable Brazilian timber sold on the international market. Most of this valuable timber has been taken from indigenous people, a fact that is spelled out in the manifesto along with other complaints about the dev-

astation caused by logging companies. The manifesto was released to newspapers, electronic bulletin boards, and conferences around the world.

1. This year marks the five hundredth anniversary of Europeans coming to the Americas. This is the right time for a deep and critical evaluation of the course followed by the societies that devolved from European contact with the peoples and nature of this region. Regarding the relationship of Brazilian society with the environment, the evaluation of these centuries can be defined as a real tragedy. During this period cycles of predatory exploitation of natural resources took place aimed at meeting foreign needs and markets. The consequences of these cycles have been always the same: environmental destruction and social impoverishment.

2. In Brazil the exploitation of Pau Brasil, a red timber and dye, was the first mark of this cruel process. Its consequences are well-known: the disorderly occupation of the territory, the cultural disintegration of indigenous populations, the destruction of forest ecosystems, the extinction of species and the dilapidation of natural resources. This took place for the benefit of a selfish elite and their international partners. (Pau Brasil became commercially extinct in the first century of colonization.)

3. Today, five centuries later, we can realize how much this pattern of exploitation, referred to as development, is still present. The Amazon region provides many examples of this pattern of development and is still submitted to an irrational process of devastation and disorderly occupation. Already 415 thousand square kilometres of the Brazilian Amazon have been deforested, about equivalent in size to Germany. The fruits of this devastation are more than questionable: the destruction of tens of indigenous cultures, huge unproductive farms generating very few jobs, illegal and highly polluting gold mining projects, colonists with abysmal living conditions and the advance of prostitution and drug dealing. Instead of recognizing these mistakes and taking up ecologically viable projects in the areas already deforested (such as agrarian reform, ecological agriculture and forest restoration) the agricultural frontier continues to push into primary forest areas reproducing the same mistakes of the past. As affirmed by the signatories to this Manifesto, colonists prefer to settle in the vast areas already cleared from the forest and do not wish to be pushed into the jungle where living conditions are much harder. It makes no sense to destroy

more virgin forest when the result of the deforestation promoted so far are totally absurd in terms of ecological and economic aspects.

4. It is ironic that one of the major economic forces that continues this destructive and disordered penetration of the Amazon forest is very similar to the activity that extinguished Pau Brasil in the beginning of the colonization of Brazil. Today, the logging industry, and especially Mahogany logging, seriously threatens the future of much of the Amazon forest.

5. Mahogany is the most valuable Brazilian timber sold on the international market. The species is found over a huge area of the southern Amazon covering much of the states of Para, Amazonas, Mato Grosso, Rondonia and Acre. The big sawmill owners who pay for the search and extraction of this "green gold" do not consider how their industry leads to the degradation and eventual complete destruction of primary forest areas. Mahogany is a rare tree and its exploitation requires loggers to move into more and more remote forest areas each year. More than 3,000 kilometres of major roads and tens of thousands of kilometres of secondary roads have been illegally pushed in the southern Amazon in the last ten years to extract Mahogany. After cutting out all marketable Mahogany in a given area the logging companies move on, leaving their access roads for colonists, gold miners and displaced poor who consolidate the destruction of the forest. The exploitation of timber in general and the cutting of Mahogany in particular is the driving force leading to forest destruction in the southern Amazon today.

6. Most of the Mahogany extracted in Brazil is taken from the territories of indigenous peoples. There are numerous reports of Mahogany trees being cut down and removed from indigenous lands despite resistance of indian communities against this usurpation of their territory. At the same time, strong pressure and the allurement of indian leaders . . . has led some indigenous communities to sign contracts allowing the extraction of Mahogany on their lands. These agreements, however, have not been approved by the relevant government authorities and are legally invalid. Such timber deals have often led to the political fragmentation and cultural degradation of indian groups. These communities learn by example that ecological destruction through over-exploitation of the forest is the only alternative that can allow them access to outside assistance and consumer goods. In the last ten years the number of indigenous communities that have become victims of the Mahogany

boom has more than doubled. The problem is bound to get worse as almost all the remaining stands of Mahogany are located inside indigenous territories. The lack of enforcement from . . . the federal environment agency, has encouraged the illegal exploitation of Mahogany from indigenous land from logging companies. Over the last two years in the state of Para, the bulk of all Mahogany produced come from trees illegally extracted from indigenous lands.

7. The ecological impacts of the Mahogany industry are equally serious. Areas of protected forest such as the Biological Reserve of Guapor in Rondonia and the Extractive Reserve Chico Mendes in Acre have been systematically invaded by Mahogany loggers. National Parks throughout the region have also been violated. Due to intense exploitation, the Amazonian Mahogany species, Sweitenia macrophylla King, was added in the official list of Brazilian species threatened with extinction in 1992. There is some debate regarding how close the species is to extinction but the fact remains that if left unchecked, the current pattern of exploitation of Mahogany will drive the species to extinction in the Amazon in a few short years. Wherever the Mahogany exploitation frontier has passed, the species has become practically extinct. In the 'sixties Mahogany was extracted in the Araguaia region of Para. After the commercial extinction of the species in this region, the loggers moved forward along the highway PA 150. During the 'eighties this same predatory pattern moved into the occidental part of Amazonia towards the Xingu River. Today the Mahogany loggers have already crossed the Xingu and keep advancing. In addition to the role that logging roads play in opening up primary forest to other destructive influences, Mahogany logging itself causes considerable ecological damage. Studies show that for each cut tree around 28 other trees are killed and some 1450 square meters of forest are damaged. The few attempts at growing the species in plantations in the Amazon are still in their early stages and only occupy a small area when compared with the amount of forest damaged by the industry each year. To date none of the plantations have demonstrated the ability to overcome attacks of the moth (Hypssipella grandella) common in commercial Mahogany plantations. Such plantations are often used to justify the continuation of the Mahogany industry in natural forests rather than a real search for concrete alternatives to forest destruction.

8. Despite all the problems mentioned above there are people and companies who still defend the exploitation of Mahogany as

a source of economic development for the Brazilian Amazon. Even this argument must be contested. The Mahogany industry is made up of an extensive chain of informal actors and middlemen who are controlled by a small elite group of sawmills and exporters. The industry generates relatively few jobs and the bulk of the profits are made in the importing countries or in the southeast of Brazil. The Mahogany sawmills belong to business groups who moved to Amazonia after exhausting the timber resources in the Atlantic Coast Rainforests and the Araucaria forests in the South of Brazil. In addition to the impunity with which Mahogany loggers still operate in nature reserves and on indigenous lands, there are strong indications that the industry includes a substantial number of companies engaging in tax fraud through concealing information regarding the source and the correct volume of extracted logs.

9. Given the seriousness of the economic, ecological and social impact of the industry and the clear evidence that Mahogany extraction is perpetuating and intensifying the chaotic model of occupation in Amazonia to the detriment of Brazilian society, the following groups, many of whom deal directly with this problem, have formulated the following demands:

A) That the Brazilian government through its competent institutions must face up to this problem by prohibiting all cutting and trade of Mahogany in the Amazon region until it has evaluated the extent of damage caused by the industry to date, and defined through an wide debate among all interested parties, legal measures necessary to halt this chaotic process. This measure is necessary to achieve the following objectives:

a) Mahogany loggers do not build illegal and inadequate roads into primary forest areas, and that existing logging roads are used appropriately or closed.

b) all exploitation of Mahogany trees in areas designated for ecological preservation is halted.

c) all exploitation of Mahogany in indigenous areas and extractive reserves is halted and at the same time the government concretely supports the efforts of the forest peoples and communities to find economic and non-predatory alternatives for their survival and development.

d) the dynamic of predatory logging leading to the extinction of species in areas reached by Mahogany exploitation is stopped.

B) That the solution of the problems caused by the Mahogany industry must be seen as the first step in the implementation of policies and programmes to end all forms of predatory logging in Amazonia. This policy on Mahogany should serve as a guide to transform all logging activities throughout the region; including restricting the areas where logging is allowed, defining rigidly the technical conditions acceptable for logging operations, halting the violation of protected areas and indigenous territories, and enforcing prohibitions on the cutting of species forbidden by law such as the Brazil-nut tree.

The Declaration of Curitiba

In March 1997, an international meeting of indigenous people and others from twenty countries opposed to destructive dams was held in Curitiba, Brazil. At the meeting, participants approved the Curitiba Declaration, which describes the devastating environmental, social, and economic effects of dam building on many forest communities while large landholders, agribusiness corporations, and speculators benefit financially. The Curitiba Declaration affirms the right to life and livelihood of people affected by dams:

We, the people from 20 countries gathered in Curitiba, Brazil, representing organizations of dam-affected people and of opponents of destructive dams, have shared our experiences of the losses we have suffered and the threats we face because of dams. Although our experiences reflect our diverse cultural, social, political and environmental realities, our struggles are one. Our struggles are one because everywhere dams force people from their homes, submerge fertile farmlands, forests and sacred places, destroy fisheries and supplies of clean water, and cause the social and cultural disintegration and economic impoverishment of our communities. Our struggles are one because everywhere there is a wide gulf between the economic and social benefits promised by dam builders and the reality of what has happened after dam construction. Dams have almost always cost more than was projected, even before including environmental and social costs. Dams have produced less electricity and irrigated less land than was promised. They have made floods even more destructive. Dams have benefited large landholders, agribusiness corporations and speculators. They have dispossessed small farmers; rural workers; fishers; tribal, indigenous and traditional communities. Our struggles are one because we are fighting against similar powerful interests, the same international lenders, the same multilateral and bilateral aid and credit agencies, the same dam construction and equipment companies, the same engineering and environmental consultants, and the same corporations involved in heavily subsidized energy-intensive industries. Our struggles are one because everywhere the people who suffer most from dams are excluded from decision-making. Decisions are instead taken by technocrats, politicians and business elites who increase their own power and wealth through building dams. Our common struggles convince us that it is both necessary and possible to bring an end to the era of destructive dams. It is also both necessary and possible to implement alternative ways of providing energy and managing our freshwaters which are equitable, sustainable and effective. For this to happen, we demand genuine democracy which includes public participation and transparency in the development and implementation of energy and water policies, along with the decentralization of political power and empowerment of local communities. We must reduce inequality through measures including equitable access to land. We also insist on the inalienable rights of communities to control and manage their water, land, forests and other resources and the right of every person to a healthy environment. We must advance to a society where human beings and nature are no longer reduced to the logic of the market, where the only value is that of commodities and the only goal profits. We must advance to a society which respects diversity, and which is based on equitable and just relations between people, regions and nations. Our shared experiences have led us to agree [to] the following:

1) We recognize and endorse the principles of the 1992 "NGO and Social Movements Declaration of Rio de Janeiro" and the 1994 "Manibeli Declaration" on World Bank funding of large dams.

2) We will oppose the construction of any dam which has not been approved by the affected people after an informed and participative decision-making process.

3) We demand that governments, international agencies and investors implement an immediate moratorium on the building of large dams until:

a. There is a halt to all forms of violence and intimidation against people affected by dams and organizations opposing dams.

b. Reparations, including the provision of adequate land, housing and social infrastructure, be negotiated with the millions of people whose livelihoods have already suffered because of dams.

c. Actions are taken to restore environments damaged by dams—even when this requires the removal of the dams.

d. Territorial rights of indigenous, tribal, semi-tribal and traditional populations affected by dams are fully respected through providing them with territories which allow them to regain their previous cultural and economic conditions—this again may require the removal of the dams.

e. An international independent commission is established to conduct a comprehensive review of all large dams financed or otherwise supported by international aid and credit agencies, and its policy conclusions implemented. The establishment and procedures of the review must be subject to the approval and monitoring of representatives of the international movement of people affected by dams.

f. Each national and regional agency which has financed or otherwise supported the building of large dams have commissioned independent comprehensive reviews of each large dam project they have funded and implemented the policy conclusions of the reviews. The reviews must be carried out with the participation of representatives of the affected people's organizations.

g. Policies on energy and freshwater are implemented which encourage the use of sustainable and appropriate technologies and management practices, using the contributions of both modern science and traditional knowledge. These policies need also to discourage waste and overconsumption and guarantee equitable access to these basic needs.

4) The process of privatization which is being imposed on countries in many parts of the world by multilateral institutions is increasing social, economic and political exclusion and injustice. We do not accept the claims that this process is a solution to corruption, inefficiency and other problems in the power and water sectors where these are under the control of the state. Our priority is democratic and effective public control and regulation of entities which provide electricity and water in a way which guarantees the needs and desires of people.

5) Over the years, we have shown our growing power. We have occupied dam sites and offices, marched in our villages

and cities, refused to leave our lands even though we have faced intimidation, violence and drowning. We have un-masked the corruption, lies and false promises of the dam in-dustry. Nationally and internationally we have worked in solidarity with others fighting against destructive develop-ment projects, and together with those fighting for human rights, social justice, and an end to environmental destruction. We are strong, diverse and united and our cause is just. We have stopped destructive dams and have forced dam builders to respect our rights. We have stopped dams in the past, and we will stop more in the future. We commit ourselves to inten-sifying the fight against destructive dams. From the villages of India, Brazil and Lesotho to the boardrooms of Washington, Tokyo and London, we will force dam builders to accept our demands. To reinforce our movement we will build and strengthen regional and international networks. To symbolise our growing unity, we declare that 14 March, the Brazilian Day of Struggles Against Dams, will from now on become the International Day of Action Against Dams and for Rivers, Water, and Life. *Águas para a vida, não para a morte!* *¡Aguas para la vida, no para la muerte!* Water for life, not for death!

Source: International Rivers Network. http://www.irn.org/programs/curitiba.html. (Reprinted with permission).

Indigenous Peoples' Declaration

When the World Trade Organization met in Seattle, Washington, from November 30 to December 3, 1999, the Indigenous Peoples' Caucus convened also. Sponsored by Indigenous Environmental Network USA/CANADA, Seventh Generation Fund USA, International Indian Treaty Council, Indigenous Peoples' Council on Biocolonialism, the Abya Yala Fund, and TEBTEBBA (Indigenous Peoples' International Centre for Policy Research and Education), the Indigenous Peoples' Caucus prepared and presented a declaration to express its members' "great concern over how the World Trade Organization is destroying Mother Earth and the cultural and biological diversity of which we are a part." The declaration spells out the complaints of indigenous people and presents proposals for possible solutions to some of the problems.

We, the Indigenous Peoples from various regions of the world, have come to Seattle to express our great concern over how the World Trade Organization is destroying Mother Earth and the cultural and biological diversity of which we are a part.

Trade liberalization and export-oriented development, which are the overriding principles and policies pushed by the WTO, are creating the most adverse impacts on the lives of Indigenous Peoples. Our inherent right to self-determination, our sovereignty as nations, and treaties and other constructive agreements which Indigenous nations and Peoples have negotiated with other nation-states, are undermined by most of the WTO Agreements. The disproportionate impact of these Agreements on our communities, whether through environmental degradation or the militarization and violence that often accompanies development projects, is serious and therefore should be addressed immediately.

The WTO Agreement on Agriculture (AOA), which promotes export competition and import liberalization, has allowed the entry of cheap agricultural products into our communities. It is causing the destruction of ecologically rational and sustainable agricultural practices of Indigenous Peoples.

Food security and the production of traditional food crops have been seriously compromised. Incidents of diabetes, cancers, and hypertension have significantly increased among Indigenous Peoples because of the scarcity of traditional foods and the dumping of junk food into our communities.

Small-scale farm production is giving way to commercial cash-crop plantations, further concentrating ancestral lands into the hands of few agri-corporations and landlords. This has led to the dislocation of scores of people from our communities, who then migrate to nearby cities and become the urban homeless and jobless.

The WTO Forests Products Agreement promotes free trade in forest products. By eliminating developed country tariffs on wood products by the year 2000, and developing country tariffs by 2003, the Agreement will result in the deforestation of many of the world's ecosystems in which Indigenous Peoples live.

Mining laws in many countries are being changed to allow free entry of foreign mining corporations, to enable them to buy and own mineral lands, and to freely displace Indigenous Peoples from their ancestral territories. These large-scale commercial mining and oil extraction activities continue to degrade our lands and fragile ecosystems, and pollute the soil, water, and air in our communities.

The appropriation of our lands and resources and the aggressive promotion of consumerist and individualistic Western culture continue to destroy traditional lifestyles and cultures. The result is not only environmental degradation but also ill health, alienation, and high levels of stress manifested in high rates of alcoholism and suicides.

The theft and patenting of our biogenetic resources is facilitated by the TRIPs (Trade-Related Aspects of Intellectual Property Rights) of the WTO. Some plants which Indigenous Peoples have discovered, cultivated, and used for food, medicine, and for sacred rituals are already patented in the United States, Japan, and Europe. A few examples of these are ayahuasca, quinoa, and sangre de drago in forests of South America; kava in the Pacific; turmeric and bitter melon in Asia. Our access and control over our biological diversity and control over our traditional knowledge and intellectual heritage are threatened by the TRIPs Agreement.

Article 27.3b of the TRIPs Agreement allows the patenting of life-forms and makes an artificial distinction between plants, animals, and micro-organisms. The distinction between "essentially biological" and "non-biological" and "microbiological" processes is also erroneous. As far as we are concerned all these are life-forms and life-creating processes which are sacred and which should not become the subject of private property ownership.

Finally, the liberalization of investments and the service sectors, which is pushed by the General Agreement of Services (GATS), reinforces the domination and monopoly control of foreign corporations over strategic parts of the economy. The World Bank and the International Monetary Fund impose conditionalities of liberalization, deregulation and privatization on countries caught in the debt trap. These conditionalities are reinforced further by the WTO.

In light of the adverse impacts and consequences of the WTO Agreements identified above, we, Indigenous Peoples present the following demands:

We urgently call for a social and environmental justice analysis which will look into the Agreements' cumulative effects on Indigenous Peoples. Indigenous Peoples should be equal participants in establishing the criteria and indicators for these analyses so that they take into consideration spiritual as well as cultural aspects.

A review of the Agreements should be done to address all

of the inequities and imbalances which adversely affect Indigenous Peoples. The proposals to address some of these are as follows;

(1) For the Agreement on Agriculture

a. It should not include in its coverage small-scale farmers who are mainly engaged in production for domestic use and sale in the local markets.

b. It should ensure the recognition and protection of rights of Indigenous Peoples to their territories and their resources, as well as their rights to continue practicing their indigenous sustainable agriculture and resource management practices and traditional livelihoods.

c. It should ensure the food security and the capacity of Indigenous Peoples to produce, consume and trade their traditional foods.

(2) With regard to the liberalization of services and investments, we recommend the following:

a. It must stop unsustainable mining, commercial planting of monocrops, dam construction, oil exploration, land conversion to golf clubs, logging, and other activities which destroy Indigenous Peoples' lands and violate the rights of indigenous peoples to their territories and resources.

b. The right of Indigenous Peoples to their traditional lifestyles, cultural norms and values should likewise be recognized and protected.

c. The liberalization of services, especially in the areas of health, should not be allowed if it will prevent Indigenous Peoples from having access to free, culturally appropriate as well as quality health services.

d. The liberalization of finance services which makes the world a global casino should be regulated.

(3) On the TRIPs Agreement, the proposals are as follows:

a. Article 27.3b should be amended to categorically disallow the patenting of life-forms. It should clearly prohibit the patenting of micro-organisms, plants, animals, including all their parts, whether they are genes, gene sequences, cells, cell lines, proteins, or seeds.

b. It should also prohibit the patenting of natural processes, whether these are biological or microbiological, involving the use of plants, animals and micro-organisms and their parts in producing variations of plants, animals and micro-organisms.

c. It should ensure the exploration and development of

alternative forms of protection outside of the dominant western intellectual property rights regime. Such alternatives must protect the knowledge and innovations and practices in agriculture, health care, and conservation of biodiversity, and should build upon indigenous methods and customary laws protecting knowledge, heritage and biological resources.

d. It should ensure that the protection offered to indigenous and traditional knowledge, innovation and practices is consistent with the Convention on Biological Diversity (i.e., Articles 8j, 10c, 17.2, and 18.4) and the International Undertaking on Plant Genetic Resources.

e. It should allow for the right of Indigenous Peoples and farmers to continue their traditional practices of saving, sharing and exchanging seeds, and cultivating, harvesting and using medicinal plants.

f. It should prohibit scientific researchers and corporations from appropriating and patenting indigenous seeds, medicinal plants, and related knowledge about these life-forms. The principles of prior informed consent and right of veto by Indigenous Peoples should be respected.

If the earlier proposals cannot be ensured, we call for the removal of the Agreement on Agriculture, the Forest Products Agreements and the TRIPs Agreement from the WTO.

General Demands: We call on the member-states of the WTO not to allow for another round whilst the review and rectification of the implementation of existing agreements has not been done. We reject the proposals for an investment treaty, competition, accelerated industrial tariffs, government procurement, and the creation of a working group on biotechnology.

We urge the WTO to reform itself to become democratic, transparent and accountable. If it fails to do this we call for the abolition of the WTO.

We urge the member nation-states of the WTO to endorse the adoption by the UN General Assembly of the current text of the UN Declaration on the Rights of Indigenous Peoples and the ratification of ILO Convention 169.

We call on the peoples' organizations and NGOs to support this "Indigenous Peoples' Seattle Declaration" and to promote it among their members.

We believe that the whole philosophy underpinning the WTO Agreements and the principles and policies it promotes contradict our core values, spirituality and worldviews, as well as our concepts and practices of development, trade and environmental protection. Therefore, we challenge the WTO to redefine its principles and practices toward a "sustainable communities" paradigm, and to recognize and allow for the continuation of other worldviews and models of development.

Indigenous peoples, undoubtedly, are the ones most adversely affected by globalization and by the WTO Agreements. However, we believe that it is also us who can offer viable alternatives to the dominant economic growth, export-oriented development model. Our sustainable lifestyles and cultures, traditional knowledge, cosmologies, spirituality, values of collectivity, reciprocity, respect and reverence for Mother Earth, are crucial in the search for a transformed society where justice, equity, and sustainability will prevail.

Source: Society for Threatened Peoples [Gesellschaft für bedrohte völker] http://www.gfbv.de/gfbv_e/docus/wto_e.htm (Reprinted with permission)

North American Free Trade Agreement and the Commission for Environmental Cooperation

As its title suggests, the North American Free Trade Agreement (NAFTA) deals with trade issues between Canada, the United States, and Mexico. A "Side Agreement to Environmental Cooperation," which was ratified by the three nations at the time of the NAFTA treaty, was designed to ensure environmental safeguards across national borders. That agreement established the Commission for Environmental Cooperation to address regional environmental concerns, which include rainforest destruction, and also to promote the effective enforcement of environmental law. Only the first two parts of this agreement are reproduced here; the entire document includes the preamble, a description of the structure and duties of the commission, provisions for cooperation between nations, guidelines for resolving disputes, rules for enforcement, and definitions of terms.

North American Agreement on Environmental Cooperation Between the Government of Canada, the Government of the United Mexican States and the Government of the United States of America (1993)

Part One—Objectives   Article 1: Objectives
The objectives of this Agreement are to:
   (a) foster the protection and improvement of the environ-

ment in the territories of the Parties for the well-being of present and future generations;

(b) promote sustainable development based on cooperation and mutually supportive environmental and economic policies;

(c) increase cooperation between the Parties to better conserve, protect, and enhance the environment, including wild flora and fauna;

(d) support the environmental goals and objectives of the NAFTA;

(e) avoid creating trade distortions or new trade barriers;

(f) strengthen cooperation on the development and improvement of environmental laws, regulations, procedures, policies and practices;

(g) enhance compliance with, and enforcement of, environmental laws and regulations;

(h) promote transparency and public participation in the development of environmental laws, regulations and policies;

(i) promote economically efficient and effective environmental measures; and

(j) promote pollution prevention policies and practices.

Part Two—Obligations Article 2: General Commitments
1. Each Party shall, with respect to its territory:

(a) periodically prepare and make publicly available reports on the state of the environment;

(b) develop and review environmental emergency preparedness measures;

(c) promote education in environmental matters, including environmental law;

(d) further scientific research and technology development in respect of environmental matters;

(e) assess, as appropriate, environmental impacts; and

(f) promote the use of economic instruments for the efficient achievement of environmental goals.

2. Each Party shall consider implementing in its law any recommendation developed by the Council under Article 10(5)(b).

3. Each Party shall consider prohibiting the export to the territories of the other Parties of a pesticide or toxic substance whose use is prohibited within the Party's territory. When a Party adopts a measure prohibiting or severely restricting the use of a pesticide or toxic substance in its territory, it shall notify the other Parties of the measure, either directly or through an appropriate international organization.

Article 3: Levels of Protection. Recognizing the right of each Party to establish its own levels of domestic environmental protection and environmental development policies and priorities, and to adopt or modify accordingly its environmental laws and regulations, each Party shall ensure that its laws and regulations provide for high levels of environmental protection and shall strive to continue to improve those laws and regulations. Article 4: Publication.

1. Each Party shall ensure that its laws, regulations, procedures and administrative rulings of general application respecting any matter covered by this Agreement are promptly published or otherwise made available in such a manner as to enable interested persons and Parties to become acquainted with them.

2. To the extent possible, each Party shall:

(a) publish in advance any such measure that it proposes to adopt; and

(b) provide interested persons and Parties a reasonable opportunity to comment on such proposed measures.

Article 5: Government Enforcement Action.

1. With the aim of achieving high levels of environmental protection and compliance with its environmental laws and regulations, each Party shall effectively enforce its environmental laws and regulations through appropriate governmental action, subject to Article 37, such as:

(a) appointing and training inspectors;

(b) monitoring compliance and investigating suspected violations, including through on-site inspections;

(c) seeking assurances of voluntary compliance and compliance agreements;

(d) publicly releasing non-compliance information;

(e) issuing bulletins or other periodic statements on enforcement procedures;

(f) promoting environmental audits;

(g) requiring record keeping and reporting;

(h) providing or encouraging mediation and arbitration services;

(i) using licenses, permits or authorizations;

(j) initiating, in a timely manner, judicial, quasi-judicial or administrative proceedings to seek appropriate sanctions or remedies for violations of its environmental laws and regulations;

(k) providing for search, seizure or detention; or

(l) issuing administrative orders, including orders of a preventative, curative or emergency nature.

2. Each party shall ensure that judicial, quasi-judicial or administrative enforcement proceedings are available under its law to sanction or remedy violations of its environmental laws and regulations.

3. Sanctions and remedies provided for a violation of a Party's environmental laws and regulations shall, as appropriate:

(a) take into consideration the nature and gravity of the violation, any economic benefit derived from the violation by the violator, the economic condition of the violator, and other relevant factors; and

(b) include compliance agreements, fines, imprisonment, injunctions, the closure of facilities, and the cost of containing or cleaning up pollution.

Article 6: Private Access to Remedies.

1. Each Party shall ensure that interested persons may request the Party's competent authorities to investigate alleged violations of its environmental laws and regulations and shall give such requests due consideration in accordance with law.

2. Each Party shall ensure that persons with a legally recognized interest under its law in a particular matter have appropriate access to administrative, quasi-judicial or judicial proceedings for the enforcement of the Party's environmental laws and regulations.

3. Private access to remedies shall include rights, in accordance with the Party's law, such as:

(a) to sue another person under that Party's jurisdiction for damages;

(b) to seek sanctions or remedies such as monetary penalties, emergency closures or orders to mitigate the consequences of violations of its environmental laws and regulations;

(c) to request the competent authorities to take appropriate action to enforce that Party's environmental laws and regulations in order to protect the environment or to avoid environmental harm; or

(d) to seek injunctions where a person suffers, or may suffer, loss, damage or injury as a result of conduct by another person under that Party's jurisdiction contrary to that Party's environmental laws and regulations or from tortious conduct.

Article 7: Procedural Guarantees.

1. Each Party shall ensure that its administrative, quasi-

judicial and judicial proceedings referred to in Articles 5(2) and 6(2) are fair, open and equitable, and to this end shall provide that such proceedings:

(a) comply with due process of law;

(b) are open to the public, except where the administration of justice otherwise requires;

(c) entitle the parties to the proceedings to support or defend their respective positions and to present information or evidence; and

(d) are not unnecessarily complicated and do not entail unreasonable charges or time limits or unwarranted delays.

2. Each Party shall provide that final decisions on the merits of the case in such proceedings are:

(a) in writing and preferably state the reasons on which the decisions are based;

(b) made available without undue delay to the parties to the proceedings and, consistent with its law, to the public; and

(c) based on information or evidence in respect of which the parties were offered the opportunity to be heard.

3. Each Party shall provide, as appropriate, that parties to such proceedings have the right, in accordance with its law, to seek review and, where warranted, correction of final decisions issued in such proceedings.

4. Each Party shall ensure that tribunals that conduct or review such proceedings are impartial and independent and do not have any substantial interest in the outcome of the matter.

Source: North American Commission on Environmental Cooperation website. http://www.cec.org/pubs_info_resources/law_treat_agree/naaec/index.cfm?varlan=english)

United Nations Documents

Two of the major documents that resulted from the "Earth Summit" held in Rio de Janeiro in June 1992 were the "Rio Declaration" and a statement of forest principles, both of which were signed by heads of state. The Rio Declaration describes the basic guidelines for nations to follow in matters of environmental protection and development. The "Statement of Forest Principles" outlines basic guidelines for the protection of all types of forests.

Report of the United Nations Conference on Environment and Development (Rio de Janeiro, 3–14 June 1992)—Annex I: Rio Declaration on Environment and Development

The United Nations Conference on Environment and Development, Having met at Rio de Janeiro from 3 to 14 June 1992, Reaffirming the Declaration of the United Nations Conference on the Human Environment, adopted at Stockholm on 16 June 1972, and seeking to build upon it, With the goal of establishing a new and equitable global partnership through the creation of new levels of cooperation among States, key sectors of societies and people, Working towards international agreements which respect the interests of all and protect the integrity of the global environmental and developmental system, Recognizing the integral and interdependent nature of the Earth, our home, Proclaims that:

Principle 1: Human beings are at the centre of concerns for sustainable development. They are entitled to a healthy and productive life in harmony with nature.

Principle 2: States have, in accordance with the Charter of the United Nations and the principles of international law, the sovereign right to exploit their own resources pursuant to their own environmental and developmental policies, and the responsibility to ensure that activities within their jurisdiction or control do not cause damage to the environment of other States or of areas beyond the limits of national jurisdiction.

Principle 3: The right to development must be fulfilled so as to equitably meet developmental and environmental needs of present and future generations.

Principle 4: In order to achieve sustainable development, environmental protection shall constitute an integral part of the development process and cannot be considered in isolation from it.

Principle 5: All States and all people shall cooperate in the essential task of eradicating poverty as an indispensable requirement for sustainable development, in order to decrease the disparities in standards of living and better meet the needs of the majority of the people of the world.

Principle 6: The special situation and needs of developing countries, particularly the least developed and those most environmentally vulnerable, shall be given special priority. International actions in the field of environment and development should also address the interests and needs of all countries.

Principle 7: States shall cooperate in a spirit of global partnership to conserve, protect and restore the health and integrity of the Earth's ecosystem. In view of the different contributions to global environmental degradation, States have common but differentiated responsibilities. The developed countries acknowledge the responsibility that they bear in the international pursuit of sustainable development in view of the pressures their societies place on the global environment and of the technologies and financial resources they command.

Principle 8: To achieve sustainable development and a higher quality of life for all people, States should reduce and eliminate unsustainable patterns of production and consumption and promote appropriate demographic policies.

Principle 9: States should cooperate to strengthen endogenous capacity-building for sustainable development by improving scientific understanding through exchanges of scientific and technological knowledge, and by enhancing the development, adaptation, diffusion and transfer of technologies, including new and innovative technologies.

Principle 10: Environmental issues are best handled with the participation of all concerned citizens, at the relevant level. At the national level, each individual shall have appropriate access to information concerning the environment that is held by public authorities, including information on hazardous materials and activities in their communities, and the opportunity to participate in decision-making processes. States shall facilitate and encourage public awareness and participation by making information widely available. Effective access to judicial and administrative proceedings, including redress and remedy, shall be provided.

Principle 11: States shall enact effective environmental legislation. Environmental standards, management objectives and priorities should reflect the environmental and developmental context to which they apply. Standards applied by some countries may be inappropriate and of unwarranted economic and social cost to other countries, in particular developing countries.

Principle 12: States should cooperate to promote a supportive and open international economic system that would lead to economic growth and sustainable development in all countries, to better address the problems of environmental degradation. Trade policy measures for environmental purposes should not constitute a means of arbitrary or unjustifiable discrimination or a disguised restriction on international trade. Unilateral actions to deal with environmental challenges outside the jurisdiction of the importing country should be avoided. Environmental measures addressing transboundary or global environmental problems should, as far as possible, be based on an international consensus.

Principle 13: States shall develop national law regarding liability and compensation for the victims of pollution and other environmental damage. States shall also cooperate in an expeditious and more determined manner to develop further international law regarding liability and compensation for adverse effects of environmental damage caused by activities within their jurisdiction or control to areas beyond their jurisdiction.

Principle 14: States should effectively cooperate to discourage or prevent the relocation and transfer to other States of any activities and substances that cause severe environmental degradation or are found to be harmful to human health.

Principle 15: In order to protect the environment, the precautionary approach shall be widely applied by States according to their capabilities. Where there are threats of serious or irreversible damage, lack of full scientific certainty shall not be used as a reason for postponing cost-effective measures to prevent environmental degradation.

Principle 16: National authorities should endeavour to promote the internalization of environmental costs and the use of economic instruments, taking into account the approach that the polluter should, in principle, bear the cost of pollution, with due regard to the public interest and without distorting international trade and investment.

Principle 17: Environmental impact assessment, as a national instrument, shall be undertaken for proposed activities that are likely to have a significant adverse impact on the environment and are subject to a decision of a competent national authority.

Principle 18: States shall immediately notify other States of any natural disasters or other emergencies that are likely to produce sudden harmful effects on the environment of those States. Every effort shall be made by the international community to help States so afflicted.

Principle 19: States shall provide prior and timely notification and relevant information to potentially affected States on activities that may have a significant adverse transboundary environmental effect and shall consult with those States at an early stage and in good faith.

Principle 20: Women have a vital role in environmental management and development. Their full participation is therefore essential to achieve sustainable development.

Principle 21: The creativity, ideals and courage of the youth of the world should be mobilized to forge a global partnership in order to achieve sustainable development and ensure a better future for all.

Principle 22: Indigenous people and their communities and other local communities have a vital role in environmental management and development because of their knowledge and traditional practices. States should recognize and duly support their identity, culture and interests and enable their effective participation in the achievement of sustainable development.

Principle 23: The environment and natural resources of people under oppression, domination and occupation shall be protected.

Principle 24: Warfare is inherently destructive of sustainable development. States shall therefore respect international law providing protection for the environment in times of armed conflict and cooperate in its further development, as necessary.

Principle 25: Peace, development and environmental protection are interdependent and indivisible.

Principle 26: States shall resolve all their environmental disputes peacefully and by appropriate means in accordance with the Charter of the United Nations.

Principle 27: States and people shall cooperate in good faith and in a spirit of partnership in the fulfillment of the principles embodied in this Declaration and in the further development of international law in the field of sustainable development.

Source: United Nations. Reprinted with permission

Report of the United Nations Conference on Environment and Development (Rio De Janeiro, 3–14 June 1992)—Annex III: Non-Legally Binding Authoritative Statement of Principles for a Global Consensus on the Management, Conservation and Sustainable Development of All Types of Forests

Preamble   (a) The subject of forests is related to the entire range of environmental and development issues and opportunities, including the right to socio-economic development on a sustainable basis.

(b) The guiding objective of these principles is to contribute to the management, conservation and sustainable development of forests and to provide for their multiple and complementary functions and uses.

(c) Forestry issues and opportunities should be examined in a holistic and balanced manner within the overall context of environment and development, taking into consideration the multiple functions and uses of forests, including traditional uses, and the likely economic and social stress when these uses are constrained or restricted, as well as the potential for development that sustainable forest management can offer.

(d) These principles reflect a first global consensus on forests. In committing themselves to the prompt implementation of these principles, countries also decide to keep them under assessment for their adequacy with regard to further international cooperation on forest issues.

(e) These principles should apply to all types of forests, both natural and planted, in all geographical regions and climatic zones, including austral, boreal, subtemperate, temperate, subtropical and tropical.

(f) All types of forests embody complex and unique ecological processes which are the basis for their present and potential capacity to provide resources to satisfy human needs as well as environmental values, and as such their sound management and conservation is of concern to the Governments of the countries to which they belong and are of value to local communities and to the environment as a whole.

(g) Forests are essential to economic development and the maintenance of all forms of life.

(h) Recognizing that the responsibility for forest management, conservation and sustainable development is in many States allocated among federal/national, state/provincial and local levels of government, each State, in accordance with its constitution and/or national legislation, should pursue these principles at the appropriate level of government.
Principles/Elements.

1. (a) States have, in accordance with the Charter of the United Nations and the principles of international law, the sovereign right to exploit their own resources pursuant to their own environmental policies and have the responsibility to ensure that activities within their jurisdiction or control do not cause damage to the environment of other States or of areas beyond the limits of national jurisdiction.

(b) The agreed full incremental cost of achieving benefits associated with forest conservation and sustainable development requires increased international cooperation and should be equitably shared by the international community.

2. (a) States have the sovereign and inalienable right to utilize, manage and develop their forests in accordance with their development needs and level of socio-economic development

and on the basis of national policies consistent with sustainable development and legislation, including the conversion of such areas for other uses within the overall socio-economic development plan and based on rational land-use policies.

(b) Forest resources and forest lands should be sustainably managed to meet the social, economic, ecological, cultural and spiritual needs of present and future generations. These needs are for forest products and services, such as wood and wood products, water, food, fodder, medicine, fuel, shelter, employment, recreation, habitats for wildlife, landscape diversity, carbon sinks and reservoirs, and for other forest products. Appropriate measures should be taken to protect forests against harmful effects of pollution, including air-borne pollution, fires, pests and diseases, in order to maintain their full multiple value.

(c) The provision of timely, reliable and accurate information on forests and forest ecosystems is essential for public understanding and informed decision-making and should be ensured.

(d) Governments should promote and provide opportunities for the participation of interested parties, including local communities and indigenous people, industries, labour, non-governmental organizations and individuals, forest dwellers and women, in the development, implementation and planning of national forest policies.

3. (a) National policies and strategies should provide a framework for increased efforts, including the development and strengthening of institutions and programmes for the management, conservation and sustainable development of forests and forest lands.

(b) International institutional arrangements, building on those organizations and mechanisms already in existence, as appropriate, should facilitate international cooperation in the field of forests.

(c) All aspects of environmental protection and social and economic development as they relate to forests and forest lands should be integrated and comprehensive.

4. The vital role of all types of forests in maintaining the ecological processes and balance at the local, national, regional and global levels through, inter alia, their role in protecting fragile ecosystems, watersheds and freshwater resources and as rich storehouses of biodiversity and biological resources and sources of genetic material for biotechnology products, as well as photosynthesis, should be recognized.

5. (a) National forest policies should recognize and duly support the identity, culture and the rights of indigenous people, their communities and other communities and forest dwellers. Appropriate conditions should be promoted for these groups to enable them to have an economic stake in forest use, perform economic activities, and achieve and maintain cultural identity and social organization, as well as adequate levels of livelihood and well-being, through, inter alia, those land tenure arrangements which serve as incentives for the sustainable management of forests.

(b) The full participation of women in all aspects of the management, conservation and sustainable development of forests should be actively promoted.

6. (a) All types of forests play an important role in meeting energy requirements through the provision of a renewable source of bio-energy, particularly in developing countries, and the demands for fuelwood for household and industrial needs should be met through sustainable forest management, afforestation and reforestation. To this end, the potential contribution of plantations of both indigenous and introduced species for the provision of both fuel and industrial wood should be recognized.

(b) National policies and programmes should take into account the relationship, where it exists, between the conservation, management and sustainable development of forests and all aspects related to the production, consumption, recycling and/or final disposal of forest products.

(c) Decisions taken on the management, conservation and sustainable development of forest resources should benefit, to the extent practicable, from a comprehensive assessment of economic and non-economic values of forest goods and services and of the environmental costs and benefits. The development and improvement of methodologies for such evaluations should be promoted.

(d) The role of planted forests and permanent agricultural crops as sustainable and environmentally sound sources of renewable energy and industrial raw material should be recognized, enhanced and promoted. Their contribution to the maintenance of ecological processes, to offsetting pressure on primary/old-growth forest and to providing regional employment and development with the adequate involvement of local inhabitants should be recognized and enhanced.

(e) Natural forests also constitute a source of goods and services, and their conservation, sustainable management and use should be promoted.

7. (a) Efforts should be made to promote a supportive international economic climate conducive to sustained and environmentally sound development of forests in all countries, which include, inter alia, the promotion of sustainable patterns of production and consumption, the eradication of poverty and the promotion of food security.

(b) Specific financial resources should be provided to developing countries with significant forest areas which establish programmes for the conservation of forests including protected natural forest areas. These resources should be directed notably to economic sectors which would stimulate economic and social substitution activities.

8. (a) Efforts should be undertaken towards the greening of the world. All countries, notably developed countries, should take positive and transparent action towards reforestation, afforestation and forest conservation, as appropriate.

(b) Efforts to maintain and increase forest cover and forest productivity should be undertaken in ecologically, economically and socially sound ways through the rehabilitation, reforestation and re-establishment of trees and forests on unproductive, degraded and deforested lands, as well as through the management of existing forest resources.

(c) The implementation of national policies and programmes aimed at forest management, conservation and sustainable development, particularly in developing countries, should be supported by international financial and technical cooperation, including through the private sector, appropriate.

(d) Sustainable forest management and use should be carried out in accordance with national development policies and priorities and on the basis of environmentally sound national guidelines. In the formulation of such guidelines, account should be taken, as appropriate and if applicable, of relevant internationally agreed methodologies and criteria.

(e) Forest management should be integrated with management of adjacent areas so as to maintain ecological balance and sustainable productivity.

(f) National policies and/or legislation aimed at management, conservation and sustainable development of forests should include the protection of ecologically viable representative or unique examples of forests, including primary/old-growth forests, cultural, spiritual, historical, religious and other unique and valued forests of national importance.

(g) Access to biological resources, including genetic material, shall be with due regard to the sovereign rights of the countries where the forests are located and to the sharing on mutually agreed terms of technology and profits from biotechnology products that are derived from these resources.

(h) National policies should ensure that environmental impact assessments should be carried out where actions are likely to have significant adverse impacts on important forest resources, and where such actions are subject to a decision of a competent national authority.

9. (a) The efforts of developing countries to strengthen the management, conservation and sustainable development of their forest resources should be supported by the international community, taking into account the importance of redressing external indebtedness, particularly where aggravated by the net transfer of resources to developed countries, as well as the problem of achieving at least the replacement value of forests through improved market access for forest products, especially processed products. In this respect, special attention should also be given to the countries undergoing the process of transition to market economies.

(b) The problems that hinder efforts to attain the conservation and sustainable use of forest resources and that stem from the lack of alternative options available to local communities, in particular the urban poor and poor rural populations who are economically and socially dependent on forests and forest resources, should be addressed by Governments and the international community.

(c) National policy formulation with respect to all types of forests should take account of the pressures and demands imposed on forest ecosystems and resources from influencing factors outside the forest sector, and intersectoral means of dealing with these pressures and demands should be sought.

10. New and additional financial resources should be provided to developing countries to enable them to sustainably manage, conserve and develop their forest resources, including through afforestation, reforestation and combating deforestation and forest and land degradation.

11. In order to enable, in particular, developing countries to enhance their endogenous capacity and to better manage, conserve and develop their forest resources, the access to and transfer of environmentally sound technologies and corresponding know-how on favourable terms, including on concessional and preferential terms, as mutually agreed, in accordance with the

relevant provisions of Agenda 21, should be promoted, facilitated and financed, as appropriate.

12. (a) Scientific research, forest inventories and assessments carried out by national institutions which take into account, where relevant, biological, physical, social and economic variables, as well as technological development and its application in the field of sustainable forest management, conservation and development, should be strengthened through effective modalities, including international cooperation. In this context, attention should also be given to research and development of sustainably harvested non-wood products.

(b) National and, where appropriate, regional and international institutional capabilities in education, training, science, technology, economics, anthropology and social aspects of forests and forest management are essential to the conservation and sustainable development of forests and should be strengthened.

(c) International exchange of information on the results of forest and forest management research and development should be enhanced and broadened, as appropriate, making full use of education and training institutions, including those in the private sector.

(d) Appropriate indigenous capacity and local knowledge regarding the conservation and sustainable development of forests should, through institutional and financial support and in collaboration with the people in the local communities concerned, be recognized, respected, recorded, developed and, as appropriate, introduced in the implementation of programmes. Benefits arising from the utilization of indigenous knowledge should therefore be equitably shared with such people.

13. (a) Trade in forest products should be based on non-discriminatory and multilaterally agreed rules and procedures consistent with international trade law and practices. In this context, open and free international trade in forest products should be facilitated.

(b) Reduction or removal of tariff barriers and impediments to the provision of better market access and better prices for higher value-added forest products and their local processing should be encouraged to enable producer countries to better conserve and manage their renewable forest resources.

(c) Incorporation of environmental costs and benefits into market forces and mechanisms, in order to achieve forest conservation and sustainable development, should be encouraged both domestically and internationally.

(d) Forest conservation and sustainable development policies should be integrated with economic, trade and other relevant policies.

(e) Fiscal, trade, industrial, transportation and other policies and practices that may lead to forest degradation should be avoided. Adequate policies, aimed at management, conservation and sustainable development of forests, including, where appropriate, incentives, should be encouraged.

14. Unilateral measures, incompatible with international obligations or agreements, to restrict and/or ban international trade in timber or other forest products should be removed or avoided, in order to attain long-term sustainable forest management.

15. Pollutants, particularly air-borne pollutants, including those responsible for acidic deposition, that are harmful to the health of forest ecosystems at the local, national, regional and global levels should be controlled.

Source: United Nations. Reprinted with permission.

6

Directory of Organizations

Thousands of local, state, regional, national, and international organizations participate in projects and programs that are designed to conserve natural resources, including temperate or tropical rainforests, with many new organizations or groups forming each year. At the same time some groups have formed in the United States to advocate for the free enterprise system and to counter what they see as the federal government's efforts to take private property in order to preserve rainforests and biodiversity.

There is space here for only a representative sampling of national and international organizations and agencies that sponsor projects to protect the global environment or that are at the forefront of what has become known as the rainforest movement. In addition, a few organizations that oppose the activities of rainforest activists are also included.

Most organizations listed here publish a variety of materials, some free and others requiring a subscription fee. Publications include books, pamphlets, magazines, newsletters, research papers, press releases, curriculum guides, and other printed matter, which are usually listed in information packets available on request. In most cases, mail inquiries about an organization can be addressed to its "director."

All of the organizations described in this chapter have websites that contain the most current information about their activities, campaigns, published materials, and other pertinent data. Although URLs are subject to change, groups usually can be located with a simple Internet search by name. Most sites list the contact information to use to make donations, ask questions, or receive further information.

Alaska Rainforest Campaign (ARC)
320 Fourth Street NE
Washington, DC 20002
Phone: 202.544.0475
Fax: 202.544.5197
http://www.akrain.org

The Alaska Rainforest Campaign (ARC) is a coalition of groups that include the Alaska Center for the Environment, Defenders of Wildlife, Earthjustice Legal Defense Fund, Alaska Conservation Foundation, Natural Resources Defense Council, Sierra Club, Southeast Alaska Conservation Council, The Wilderness Society, Alaska Wilderness League, Anchorage Audubon Society, Sitka Conservation Society, and The Boat Company. The coalition works to protect the remaining wild lands of the Alaskan Tongass and Chugach National Forests from clear-cutting and other harmful development. One of the coalition's major concerns is that the U.S. Forest Service has recommended excluding the Tongass—the nation's largest and wildest national forest—from a plan to protect the remaining roadless wildlands of our national forests. Any action on this issue has been deferred until 2004, when the Forest Service plans to do another study of the forest; however, the Forest Service has made no commitment to provide any additional protection for the Tongass. ARC wants to prevent the addition of 400 miles of logging roads to the more than 4,000 miles of roads that already snake through the Tongass rainforest. But the Alaskan members of the U.S. Congress and their supporters contend that logging in the Tongass is needed to bolster the economy of the region.

The ARC website includes information about the coalition, its action alerts, and what people can do to help with campaigns, as well as general information on rainforests.

American Forest and Paper Association (AF&PA)
1111 Nineteenth Street, NW, Suite 800
Washington, DC 20036
Phone: 202.463.2445
http://www.afandpa.org/

The American Forest and Paper Association (AF&PA) is the national trade association of the forest, paper, and wood products industry. According to AF&PA's website, the association represents "member companies engaged in growing, harvesting, and processing wood and wood fiber, manufacturing pulp, paper, and paperboard products from both virgin and recycled fiber, and producing engineered and traditional wood products." AF&PA industry members account for over 8 percent of the total U.S. manufacturing output.

The AF&PA website includes information about current federal and state legislation pertaining to forests and the organization's position on climate change, a history of papermaking and other relevant facts, and information about paper and wood recycling.

American Forests (AF)
P.O. Box 2000
Washington, DC 20013
Phone: 202.955.4500
http://www.americanforests.org/

American Forests (AF) was founded in 1875 to maintain and improve the health and value of trees and forests. Through publications and action programs, the organization attempts to interest individuals, industries, and governments in forest conservation. AF sponsors Global ReLeaf, a major tree-planting effort with projects under way worldwide to improve the Earth's environment.

Information about Global ReLeaf is available on the AF website. In addition, the site includes access to data about climate change, information about tree planting, news and publications about forest conservation, and ways to contact and support AF.

Association of Forest Service Employees
for Environmental Ethics (AFSEEE)
P.O. Box 11615
Eugene, OR 97440
Phone: 541.484.2692
http://www.afseee.org/

After twelve years of growing frustration with the U.S. Forest Service (USFS) practices, Jeff DeBonis, a timber sale planner, began organizing fellow employees to speak out against agency mismanagement. Thousands of present, retired, and former USFS employees responded, and the Association of Forest Service Employees for Environmental Ethics (AFSEEE) was formed. Its purpose, stated on its website, is to "forge a socially responsible value system for the Forest Service based on a land ethic which ensures ecologically and economically sustainable management."

Members of AFSEE believe that management of the U.S. forests should be based on biological diversity, resource sustainability, and the integrity of interrelated ecosystems rather than commodity production and political expediency. The membership asserts that while timber cutting, grazing, mining, and other resource management activities can be appropriate use of public lands, these activities should not compromise other resource options and values.

On its website, AFSEEE describes its mission and projects, explains how to join the group, and provides a guide to free speech in the USFS workplace. The site also includes a middle school curriculum about forests, news items about the USFS, and ways the public can influence forest policy in the United States.

Canadian Rainforest Network (CRN)
Box 2241
Main Post Office
Vancouver, BC, Canada
V6B 3W2
Phone: 604.669.4303
Fax: 604.669.6833

The Canadian Rainforest Network (CRN) was formed in 1996 by a coalition of conservation groups in British Columbia, Canada. Because of the rapidly disappearing coastal rainforests, the CRN says it promotes "fundamental changes in the way British Columbians approach both forestry and forest protection," developing conservation strategies that "will ensure the continued ecological integrity of BC's rainforest and the communities that depend on it." The CRN works to gain permanent protection for key ecological areas in the temperate rainforests, to stop all clearcutting, to halt road building in key ecological areas, to help safeguard the traditional territories and cultures of First Nations, and to promote ecologically responsible community development.

Center for Conservation Biology (CCB)
Department of Biological Sciences
Stanford University
Stanford, CA 94305–5020
Phone: 650.723.5924
Fax: 650.723.5920
http://www.stanford.edu/group/CCB/

Founded in 1984, the Center for Conservation Biology (CCB) is an independent research group associated with Stanford University. Well-known Stanford professor of population studies Paul Ehrlich is president of the organization, which focuses on conservation biology, a science that combines the study of population biology and ecology, applying scientific techniques to manage populations and ecosystems. One of the CCB's goals is to prevent further loss to species and habitats.

The CCB's mission is described on its website, along with information about current projects, staff, and resources. A "connections" site is designed to help the media connect environmental events, causes, and effects.

Center for the Defense of Free Enterprise (CDFE)
Liberty Park
12500 NE Tenth Place
Bellevue, WA 98005
Phone: 425.455.5038
Fax: 425.451.3959
http://www.cdfe.org/

According to its website, the Center for the Defense of Free Enterprise (CDFE) was founded in 1976 by a group "deeply concerned about the rollback of 200 years of individual rights and the multitude of restrictions being imposed on America's free enterprise system by big government—and the lack of understanding of this problem by the American people." The CDFE informs and advocates for free enterprise through its publishing house, a legal defense fund, guest columns in newspapers and magazines, an investigative task force, and community organizing.

The CDFE mission and descriptions of its projects are included on its website.

Center for International Environmental Law (CIEL)
1367 Connecticut Avenue NW, Suite #300
Washington, DC 20036
Phone: 202.785.8700
Fax: 202.785.8701
http://www.ciel.org

The Center for International Environmental Law (CIEL) is a nonprofit environmental law firm founded in 1989 to strengthen international and comparative environmental law and policy around the world. On its website, CIEL declares that the center "provides a full range of environmental legal services in both international and comparative national law, including: policy research and publication, advice and advocacy, education and training, and institution building." The center works on a variety of environmental projects, including its Climate Change Program, which is designed to help prevent the threat of global warming.

Another CIEL effort is its Biodiversity and Wildlife Program to promote the "development and enforcement of an international framework of law and policy that supports conservation and sustainable use of living resources, including biodiversity, wildlife, forests and fisheries." One part of that program is to strengthen international legal mechanisms that reward local conservation efforts. Another effort is to "promote effective implementation of international conservation agreements."

Other programs and topics described on the CIEL website include human rights and the environment and trade and the environment. Its homepage also provides links to environmental law sites and publications.

Conservation International (CI)
2501 M Street NW, Suite 200
Washington, DC 20037
Phone: 202.429.5660
Toll-Free: 800.429.5660
Fax: 202.887.0193
http://www.conservation.org/

Organized in 1987, Conservation International (CI) declares that it is dedicated to the preservation of the "world's most endangered biodiversity through scientific programs, local awareness campaigns, and economic initiatives." CI initiated the first debt-for-nature swaps and works in partnership with indigenous peoples to sustain biological diversity and the ecological processes that support life on earth. CI recognizes that "cultural diversity and biological diversity are two sides of the same leaf," and works with indigenous groups to protect their traditional territories. In addition, according to the organization's website, CI "works with multinational institutions, provides economic analyses for national leaders, and promotes 'best practices' that allow for sustainable development."

On its website, CI offers access to its Ecotravel Center, which provides information about travel opportunities that benefit the environment and people. The site also includes numerous descriptions about the places where CI works, what the organization does, and how to support its efforts.

Cultural Survival (CS)
215 Prospect Street
Cambridge, MA 02139
Phone: 617.441.5400
Fax: 617.441.5417
http://www.cs.org/

Established in 1972, Cultural Survival (CS) supports projects that help isolated societies take control of their own destinies and survive contacts with the outside world, which often bring disease, relocation, impoverishment, and even death to indigenous people. CS provides funds and expertise for projects that are implemented by tribal people and ethnic minorities around the world. CS resource management programs enhance the ability of native peoples to manage natural resources. Through its nonprofit marketing division, Cultural Survival Enterprises, the organization establishes markets for products that native communities can harvest in a sustainable manner. A catalog of products is available on request.

The CS website features articles from its publication, *Cultural Survival Quarterly*, books, educational resources, the perspectives of indigenous leaders, information about special projects, and contact names and email addresses.

Defenders of Wildlife (DW)
1101 Fourteenth Street NW, Suite 1400
Washington, DC 20005
Phone: 202.682.9400
Fax: 202.682.1331
http://www.defenders.org/

According to its website, the mission of Defenders of Wildlife (DW) is to protect "native wild animals and plants in their natural communities. We focus our programs on what scientists consider two of the most serious environmental threats to the planet: the accelerating rate of extinction of species and the associated loss of biological diversity, and habitat alteration and destruction." The organization "also advocates new approaches to wildlife conservation that will help keep species from becoming endangered."

The DW website provides information about the organization, how to donate and get involved, news about DW's activities, press contacts, and wildlife programs and publications.

Earth Island Institute (EII)
300 Broadway, Suite 28
San Francisco, CA 94133
Phone: 415.788.3666
Fax: 415.788.7324
http://www.earthisland.org/

Founded in 1982 by David Brower—the Sierra Club's first executive director and the founder of Friends of the Earth and the League of Conservation Voters—Earth Island Institute (EII) supports international action projects to protect and restore the environment. EII's more than thirty projects include educating the public about global warming; coordinating conferences on the interrelationships between environmental protection, economic development, and human rights; informing the public about threats to old-growth forests in the Pacific Northwest; and providing educational materials on protecting rainforests, marine mammals, sea turtles, and indigenous lands. The organization also promotes organic and sustainable agriculture, ecological alternatives to paper, and community-based habitat restoration.

Feature articles from the *Earth Island Journal* are available on the EII website, which includes a newsroom, bookstore, and a project directory, and information on membership and support.

Earthwatch Institute (EI)
3 Clock Tower Place, Suite 100
Box 75
Maynard, MA 01754
Phone: 978.461.0081
Toll-Free: 800.776.0188
Fax: 978.461.2332
http://www.earthwatch.org/

Earthwatch (EW) is an international nonprofit organization that was established in Boston in 1971 to promote the sustainable conservation of natural resources and cultural heritage by creating partnerships between scientists, educators, and the general public. EW volunteers help scientists on research expeditions around the world, sharing in the cost of the expeditions.

The organization believes that "teaching and promoting scientific literacy is the best way to systematically approach and solve the many complex environmental and social issues facing society today." Acting as "a catalyst and a liaison between the scientific community, conservation and environmental organizations, policy makers, business, and the general public," EW conducts scientific research projects to sustain the environment in more than fifty countries worldwide. A number of projects have concentrated on vanishing rainforests and threatened species and their habitats.

Expeditions needing volunteers are described on the organization's website. Information about membership, conferences, and other events also are included.

Ecotrust
1200 NW Naito Parkway, Suite 470
Portland, OR 97209
Phone: 503.227.6225
Fax: 503.222.1517
http://www.ecotrust.org/

A nonprofit organization, Ecotrust was established in 1991 in Portland, Oregon, and is dedicated to building a conservation economy in North America's Pacific Northwest, the region from San Francisco, California, to Anchorage, Alaska, where coastal temperate rainforests are located. Working in both urban and rural areas, Ecotrust supports conservation entrepreneurs whose work improves environmental, economic, and social conditions. The organization helps local people understand the rainforest ecosystems and assists with local efforts to develop ecologically and economically sound communities in North American rainforests.

Ecotrust believes that its work is vital to other areas of North America as well. Its mission, programs, publications, and email address are available on its website.

Environmental Defense (ED)
257 Park Avenue South
New York, NY 10010
Phone: 212.505.2100
Toll-Free: 800.684.3322
Fax: 212.505.2375
http://www.edf.org/

An organization of lawyers, scientists, and economists, Environmental Defense (ED, formerly the Environmental Defense Fund) was founded in 1967. Among its projects, ED supports an online global warming exhibition called Focus on the Future and a website for the Alliance for Environmental Innovation, a joint project with The Pew Charitable Trusts that works with private companies to create environmental change. For years, ED pressured for changes in the practices of the World Bank and other development banks that lent funds for commercial logging operations in pristine rainforests. In 1991, the World Bank pledged not to fund such projects. ED also maintains an Action Network to influence national environmental policy, working with grassroots groups at the local and regional level in the United States and abroad.

ED publishes a free email newsletter, available on request; a subscription form is provided on its website. Also on the site is access to a list of ED publications and reports, a kids' page, membership information, and descriptions of ED activities.

**Food and Agriculture Organization
of the United Nations (FAO)**
Via Terme di Caaracalla
I.00100 Rome, Italy
Phone: 39.6.57971
Liaison Office for North America:
1001 Twenty-second Street NW
Washington, DC 20437
Phone: 202.653.2402
http://www.fao.org

The Food and Agriculture Organization (FAO) of the United Nations was founded in 1945 with a mandate to raise levels of nutrition and standards of living, to improve agricultural productivity, and to better the condition of rural populations. Today, FAO is the largest autonomous agency within the UN system.

Forestry has been an important part of FAO's mandate to reduce food insecurity and rural poverty, and its Forestry Department develops policies and provides advisory and technical services to FAO members regarding effective conservation and sustainable management of forests and forest resources. After consultations with member countries over a three-year period beginning in 1996, the FAO developed a Strategic Plan for Forestry that covers a long-range period from 2000 to 2015. The document, which will continually be developed and adapted to meet changing circumstances, is available on the FAO website.

The site, which is accessible in several languages, includes access to statistical databases and articles and news about agriculture, economics, fishery, forestry, nutrition, sustainable development, and many other topics.

Forest History Society (FHS)
701 William Vickers Avenue
Durham, NC 27701–3162
Phone: 919.682.9319
http://www.lib.duke.edu/forest/

Affiliated with Duke University, the Forest History Society (FHS) was established in 1946 and became a nonprofit educational institution in 1955. The society says it links "the past to the future by identifying, collecting, preserving, interpreting, and disseminating information on the history of interactions between people, forests, and their related resources—timber, water, soil, forage, fish and wildlife, recreation, and scenic or spiritual values." Focusing on North American issues within a global context, the society maintains historical records related to forests. It also maintains computer databases on forests and provides the information to scholars and others researching forest history. In addition, FHS has produced several films on forestry, such as *Timber on the Move: A History of Log Moving Technology*, and has published oral history interviews with former chiefs of the U.S. Forest Service and USFS scientists and planners.

These interviews, plus contents of the current and past issues of the scholarly journal, *Environmental History*, are available on the FHS website, which also includes a forest history bibliography, a history of the U.S. Forest Service, and library archives.

Forest Stewardship Council (FSC)
Avenida Hidalgo 502
68000 Oaxaca, México
Phone: 52.951.46905, 63244
Fax: 52.951.62110
http://www.fscoax.org/

Founded in 1993, the Forest Stewardship Council (FSC) is an international nonprofit organization that "supports environmentally appropriate, socially beneficial, and economically viable management of the world's forests." Its members represent environmental and community forest groups, the timber industry, forestry professionals, indigenous people's organizations, and forest product certification organizations worldwide.

From headquarters in Oaxaca, Mexico, the FSC focuses on clearing up the confusion about the many different environmental claims for forest products. The council has developed an international labeling scheme, which "provides a credible guarantee that the product comes from a well managed forest." According to the FSC, "All forest products carrying [their] logo have been independently certified as coming from forests that meet the internationally recognized FSC Principles and Criteria of Forest Stewardship. In this way FSC provides an incentive in the market place for good forest stewardship. The forest inspections are carried out by a number of FSC accredited certification bodies, which are evaluated and monitored to ensure their competence and credibility."

FSC guidelines for certification are included on its website, along with its mission statement, information on what the council does, and how to get involved.

Friends of the Earth (FoE)
1025 Vermont Avenue NW
Washington, DC 20005
Phone: 877.843.8687 or 202.783.7400
Fax: 202.783.0444
http://www.foe.org/

Friends of the Earth (FoE), founded in 1969, merged in 1990 with the Environmental Policy Institute and the Oceanic Society, forming a single organization, Friends of the Earth. FoE states that it "is an independent, global advocacy organization that works at local, national, and international levels to: protect the planet; preserve biological, cultural, and ethnic diversity; and empower citizens to have a voice in decisions affecting their environment and lives." Among the many projects FoE supports are national and international efforts to stop tropical deforestation and promote forest preservation. Its Green Scissors campaign is an alliance of environmentalists and conservative taxpayer organizations that aims to cut government subsidies for environmental destruction; a yearly Green Scissors report is available on the FoE website.

Also available on the website are reports of various FoE international programs to protect the global environment, descriptions of FoE publications that can be downloaded, and reports on activities of the UN Environment Programme and other UN agencies whose decisions can have an impact on the environment.

Fundación Maquipucuna/Maquipucuna Foundation (FM)
Ecuador Address:
Baquerizo #238 y Tamayo
P.O. Box 17.12.167
Phone: 593.2.507200.202

Fax: 593.2.507201
U.S. Address:
240 Parthenon Lane, Suite 1
Athens, GA 30605–2924
Phone: 706.369.9019
Fax: 706.208.9655
http://www.arches.uga.edu/~maqui/

Fundación Maquipucuna (FM), a nonprofit, nongovernmental nature conservation organization, was formed in 1988 by a multidisciplinary group of Ecuadorian citizens. Since then, the organization's mission has been to lead efforts to integrate biodiversity conservation with the rational management of natural resources in Ecuador. FM supports private land purchases for conservation, local participation in natural resource management, and the development of financial and administrative self-sufficiency. With the help of The Nature Conservancy's international program, FM bought 7,410 acres (3,000 hectares) of primary rainforest land to establish the Maquipucuna Nature Reserve in 1988, thus protecting one of the last remaining cloud forests in northwestern Ecuador from rapid deforestation due to timber, agriculture, cattle, and charcoal production. With subsequent purchases of adjacent abandoned farms, the reserve now consists of more than 9,800 acres (4,000 hectares). Ecotourism and fund-raising efforts in the United States, Great Britain, and other countries help provide financial support for the reserve.

A description of the reserve, of the available facilities, and fees and activities are on the foundation's website along with information about the organization and how to support it.

Greenpeace USA
702 H Street NW
Washington, DC 20001
Phone: 800.326.0959
http://www.greenpeaceusa.org

Greenpeace began in 1971 in protest against nuclear testing in Alaska and has expanded to become an international organization known for its nonviolent, direct-action campaigns to prevent environmental pollution. Today's campaigns focus on threats to air, water, land, and temperate and tropical rainforest resources worldwide.

In 1998, Greenpeace established a permanent base in Manaus, Brazil, and in May 2000 it launched the ship the *Amazon*

Guardian on a four-month tour of the Amazon as part of a major effort to document destruction of the Amazon rainforest and to expose illegal logging operations. According to Greenpeace, "Almost 80 percent of logging in the Amazon is illegal. Currently Greenpeace is working with indigenous communities and local groups to help find a way of maintaining the delicate ecological balance of the Amazon . . .[and] offer alternative forms of employment and opportunities for people in the region to make a living from the forest without destroying it. This includes extraction of nontimber forest products, eco-tourism, and ecologically sustainable logging."

Pictures and a diary of the tour are available on the Greenpeace website. Other Greenpeace campaigns, such as saving ancient forests in British Columbia, are also described on the website, along with information about Greenpeace publications, its online store, archives, news items, and membership forms.

International Union for the Conservation of Nature (IUCN) or the World Conservation Union (WCU)
Rue Mauverney 28
1196 Gland
Switzerland
Director General's Office
Phone: 41.22.999.0152
Fax: 41.22.999.0015
http://www.iucn.org/

The International Union for the Conservation of Nature (IUCN), known as the World Conservation Union, and was created in 1948 to bring together government agencies of many nations, nongovernmental organizations, and thousands of scientists and other experts to encourage the conservation of natural resources worldwide. IUCN's mission "is to influence, encourage and assist societies throughout the world to conserve the integrity and diversity of nature and to ensure that any use of natural resources is equitable and ecologically sustainable." Through its worldwide network, IUCN has helped countries implement national conservation strategies and protect their endangered species.

One of IUCN's projects is the "Red List" of Threatened Species, which currently identifies more than 18,000 animals and plants at risk of extinction. The Red List has been published as a booklet, produced on a CD-ROM, and is posted on the IUCN website. The site also includes species information, descriptions of "ecospaces," and outlines of IUCN's many and varied programs.

International Rivers Network (IRN)
1847 Berkeley Way
Berkeley, CA 94703
Phone: 510.848.1155
Fax: 510.848.1008
http://irn.org/

Working with communities around the world, the International Rivers Network (IRN) was established in 1985 to promote the concept that rivers and their watersheds are living systems that need to be protected and nurtured for the benefit of the human and biological communities that depend on them. Part of IRN's mission is to "halt and reverse the degradation of river systems; to support local communities in protecting and restoring the well-being of the people, cultures and ecosystems that depend on rivers; and to promote sustainable, environmentally sound alternatives to damming and channeling rivers." Some IRN projects focus on preventing the construction of dams that destroy indigenous rainforest communities by flooding their homes.

The IRN website includes descriptions of its campaigns, publications such as reports and books, videos and slides, and membership information.

**International Society for the Preservation
of the Tropical Rainforest (ISPTR)**
3302 N. Burton Avenue
Rosemead, CA 91770
Phone: 626.572.0233
Fax: 626.572.9521
http://www.isptr-pard.org/index.html

The International Society for the Preservation of the Tropical Rainforest (ISPTR) is an American and Peruvian nonprofit organization comprised of volunteer naturalists and environmentalists. Since the early 1980s, this dedicated team has been working on the front line of the Amazon Basin, setting up protected community reserves and wildlife sanctuaries and implementing a series of innovative pilot projects for the benefit of the indigenous people and the fauna and flora of the tropical forest.

In 1990, ISPTR acquired a 1,000-acre research site on the Yarapa River, a tributary of the Amazon. Scientists at the research station conduct multidisciplinary research on tropical rainforest ecosystems. Through an agreement with the National University of the Peruvian Amazon in Iquitos, Florida International University, and Ohio State University, accredited courses and exchange programs are offered at the research camp, which includes a "living classroom." The station also has established an outpost medical clinic for people who live on the river and a rehabilitation center where injured, diseased, and orphaned animals are nursed back to health and then released to safe, natural habitats. In addition, ISPTR promotes carefully managed educational and conservational ecotourism, offering trips to the Peruvian Amazon rainforest and "hands-on" experience at the noninvasive research camp.

Another project, which is featured on the ISPTR website, is the Preservation of the Amazonian River Dolphin. Photos of the rare *Inia geoffrensis* (pink dolphins) that this project aims to protect are included on the site. Links to other Amazon rainforest resources and contact information are also provided.

International Tropical Timber Organization (ITTO)
International Organizations Center, 5th Floor
Pacifico, Yokohama
1.1.1, Minato.Mirai, Nishi.ku,
Yokohama, 220.0012 Japan
Phone: 81.45.223.1110
Fax: 81.45.223.1111
http://www.itto.or.jp/Index.html

In late 1986, in fulfillment of the International Tropical Timber Agreement (1983), the International Tropical Timber Organization (ITTO) was established in Yokohama, Japan. Although created under the auspices of the United Nations, the ITTO is an independent intergovernmental organization, operating outside the UN system. Members consist of twenty-two developing nations that have control of more than 70 percent of the Earth's tropical forests and twenty-four consuming, primarily industrialized, countries. Although one of ITTO's functions is to promote trade in timber and other forest products, the organization also promotes research in sustainable timber management and national policies that encourage sustainable development.

The ITTO website offers tropical timber market reports, timber policy developments, documents from ITTO committee meetings, a newsletter, and annual reports, along with contact information.

LightHawk
The Presidio
Building 1007
P.O. Box 29231
San Francisco, CA 94129
Phone: 415.561.6250
Fax: 415.561.6251
http://www.lighthawk.org/

Known as the "Environmental Air Force," LightHawk came about in 1979 because of bush pilot Michael Stewartt's idea to let others get an aerial view of the environmental damage resulting from industrialization and deforestation. Stewartt believed that an aerial view of environmental degradation was far more convincing than pages of facts and figures attempting to prove the same thing. Now more than 150 pilots volunteer their time and their aircraft to take government officials, leaders of environmental groups, journalists, and others on flights over endangered wildlands, including rainforests, in the Western Hemisphere.

Because of LightHawk flights, the public and policy makers have seen for themselves the kind of devastation that is not visible from the ground. In flights over forests in the Pacific Northwest, for example, officials found that the U.S. Forest Service vastly overestimated the amount of ancient forest left in national forests. Huge clear-cut patches are visible from the air but do not appear on Forest Service maps. In the same way, LightHawk flights over British Columbia are designed to raise awareness of the threatened coastal rainforests along Clayoquot Sound, on Vancouver Island, and in the remote mainland coastal rainforests. LightHawk flies loggers, fishermen, and other community members and leaders over the area to give them what the organization describes as "an unbiased view at the devastating industrial scale clearcutting around Vancouver Island (it is 74 percent clearcut), as well as the unparalleled beauty and ecological importance of Clayoquot Sound and its surrounding watersheds."

LightHawk flights have helped to spot and stop destructive and illegal logging practices in Costa Rican rainforests, and they have helped prompt land conservation projects in Belize, Guatemala, and Costa Rica. The organization's Central American Program includes flights over many remote, wild areas in Honduras that otherwise would be difficult to monitor and protect. In addition, flights take place over Mexico, which struggles to promote environmental awareness and to protect the Yucatan and its endangered species.

The see-for-yourself concept has helped many environmental groups trying to protect temperate or tropical rainforests, and LightHawk has been instrumental in helping to shape public opinion about the need for conservation of rainforests and other natural resources.

The LightHawk website includes descriptions of its programs, flight operations, and current issues, a history of the organization, and an ever-growing image library that eventually will contain thousands of photographs. Contact information is also available on the site.

National Audubon Society (NAS)
1901 Pennsylvania Avenue NW
Washington, DC 20006
Phone: 202.861.2242
Fax: 202.861.4290
http://www.audubon.org/

Founded in 1905 by John James Audubon (1785–1851), famed ornithologist, explorer, and wildlife artist, the National Audubon Society is one of the oldest conservation organizations in the United States, conducting numerous campaigns to protect wildlife, wetlands, and forests. John Audubon understood the link between birds and their forested habitat and predicted early on that passenger pigeons, which relied on forests for food and shelter, would disappear as forests were clear-cut; of course, eventually the birds did become extinct.

One of the major campaigns of NAS, the Forest Habitat Campaign, is an effort to preserve America's endangered forests and the habitats vital to various species, including fish species such as Coho salmon and bull trout, which are threatened by logging that pollutes streams and spawning beds. According to NAS, "The greatest threat to these and unknown other species, and to the whole, interconnected forest ecosystem, may be the disruption of evolutionary processes. Trees we know are renewable, but we can't say the same for forests. We don't know, for example, what parts of the forest web are no longer present in a re-grown forest. These losses may mount after we cut an area three, then four times, especially if we control what species are allowed to grow. Forests may become less resilient over time, losing their ability to bounce back, to adapt to changing conditions. We like to think that forests can overcome any setback, and perhaps they can, but we can't know for sure. Even if forests do have a perpetual ability to restore themselves, the time it takes might be too long for sensitive species to wait, or too long for humans to endure." The long-suffering forests of Europe, and their native species, have not exhibited the hoped for resiliency: many of the species have never returned, and many of the forests appear to be incurably sick." NAS is also involved in a variety of international conservation efforts.

National Wildlife Federation (NWF)
8925 Leesburg Pike
Vienna, VA 22184
Phone: 703.790.4000
http://www.nwf.org/

Founded in 1936, the National Wildlife Federation (NWF) is dedicated to proper management of natural resources and sponsors numerous conservation and environmental education programs. NEF says its mission is "to educate, inspire and assist individuals and organizations of diverse cultures to conserve wildlife and other natural resources and to protect the Earth's environment in order to achieve a peaceful, equitable and sustainable future." In its "Keep the Wild Alive" campaign, NWF calls attention to and works to protect such endangered species as the rosy periwinkle, threatened by the loss of its rainforest habitat on the African island of Madagascar, and the Chinook salmon of the Pacific Northwest, which are rapidly decreasing in the wild because of the effects of dams, river silt, and "pollution from nearby logging, mining, ranching, agriculture, and development."

These and other campaigns are described on NWF's website. Its environmental education programs, publications, and ways to support NWF efforts are explained on the site as well.

Natural Resources Defense Council (NRDC)
40 West Twentieth Street
New York, NY 10011
Phone: 212.727.2700
Fax: 212.727.1773
http://www.nrdc.org/

Since the Natural Resources Defense Council (NRDC) was first initiated by several law school graduates in 1970, the organization has worked to protect endangered natural resources in the Americas and to improve the human environment through a combination of scientific research, legal action, and public education. The NRDC frequently provides legal assistance and representation to various environmental causes, such as protecting rainforests and banning the use of pesticides that pose health threats. Current projects include a campaign to protect the Great Bear rainforest in British Columbia and the endangered rainforest and the ancient forests of the Pacific Northwest in the United States.

From NRDC's homepage there are links to its sites on clean air and energy; global warming; clean water and oceans; wildlife and fish; parks, forests, and wildlands; toxic chemicals and health; nuclear weapons and waste; cities and green living; and environmental legislation. The organization's publications and contact information are also available on the site.

New Forests Project (NFP)
International Center
731 Eighth Street SE
Washington, DC 20003
Phone: 202.547.3800
Fax: 202.546.4784
http://www.newforestsproject.com/

The nonprofit International Center, which was founded in 1977 and promotes democratic movements in developing countries, established the New Forests Project (NFP) in 1982 to help curb deforestation in developing nations and provide means for restoring degraded soil caused by deforestation. With the emphasis on self-help, NFP has set up demonstration sites where field-workers show local people how to plant, wisely manage, harvest, and use fast-growing trees.

Since 1982, NFP has helped farmers begin small-scale reforestation and agroforestry projects in more than 4,000 villages in over 130 developing countries. According to NFP, the program focuses on "fast-growing, nitrogen-fixing tree species that have the potential for multiple uses. With proper management, these trees can sustainably produce fuelwood, livestock feed, and organic fertilizer. In addition to being excellent choices for soil rehabilitation and erosion control projects, these trees have a proven ability to increase agricultural yields."

NFP is described fully on its website, which also includes information on the World Seed Program, nitrogen-fixing tree species, and agroforestry.

The Nature Conservancy (TNC)
4245 North FairFax Drive, Suite 100

Arlington, VA 22203–1606
Phone: 800.628.6860
http://www.tnc.org

Founded in 1951, The Nature Conservancy (TNC) preserves habitats and species by buying and saving the lands and waters they need to survive, managing more than 1,300 preserves worldwide. TNC works with local people, creating partnerships with landowners, corporations, and governments. Some of TNC's projects involve rainforest protection. For example, in Costa Rica "The Conservancy and the Talamanca Corridor Commission are creating a continuous protected forest that extends from the top of the Continental Divide down to nine miles of undeveloped coastline along the Caribbean. The Corridor, encompassing almost 78,000 acres, connects La Amistad International Park to the coast and includes Hitoy Cerere Biological Reserve, Gandoca-Manzanillo Wildlife Refuge, several indigenous reserves, and numerous private lands. More than 90 percent of Costa Rica's known plant species are found in the Corridor," according to the TNC.

TNC also has helped establish tropical rainforest reserves in other Latin American countries, including Peru, Brazil, and Colombia. Other nature preserves managed by TNC are found in Asia and the Pacific, the Caribbean, Canada, and the United States.

These efforts and others are explained on the TNC website, along with general information about the organization, conservation science, and how people can support conservation efforts.

Oregon Natural Resources Council Fund (ONRC)
5825 North Greeley
Portland, OR 97217–4145
Phone 503.283.6343
Fax: 503.283.0756
http://www.onrc.org/

Founded in 1973, the Oregon Natural Resources Council (ONRC) is a coalition of conservation, recreation, commercial, educational, and sports groups concerned about the wise management of Oregon's natural resources. In 1995, ONRC changed its name to ONRC Fund. Although a state organization, its efforts have broad impact. Leaders of the ONRC lobby U.S. Congressional members to enact legislation that will protect the ancient forests of the Pacific Northwest and to ban exports of timber from old-growth forests.

According to its website, "ONRC was the first group to appeal a [U.S.] Forest Service timber sale based on concern for the spotted owl, and . . . helped mastermind . . . the judicial decision that halted virtually all logging in spotted owl habitat managed by the federal government. ONRC was also instrumental in [the enactment of] . . . the Endangered American Wilderness Act and the Oregon Wilderness Act, which together protect more than 1.2 million acres of Oregon's most threatened wild lands."

The ONRC website contains information on the Adopt-A-Wilderness program, volunteer opportunities, letter-writing alerts, current events, links to conservation and government sites, and ways to donate and take action.

The Pachamama Alliance (PA)
P.O. Box 29191
Presidio Bldg. #1007, Suite 215
San Francisco, CA 94129–9191
Phone: 415.561.4522
Fax: 415.561.4521
http://www.pachamama.org/

The Pachamama Alliance (PA) is a nonprofit organization, based in the United States, that works with the Achuar, a remote indigenous group of about 3,500 living in the Amazon basin in southeastern Ecuador. The Achuar make their home in nearly 2 million acres of pristine tropical rainforest in one of the most biologically diverse regions of the world. According to PA, its relationship with the Achuar started in 1995 and was "initiated by the indigenous elders and shamans themselves who, out of their deep concern for the growing threat to their ancient way of life, and their recognition that the roots of this threat lay far beyond their rainforest home, actively sought the partnership of committed individuals living in the modern world." PA was publicly launched 1997 in San Francisco, California.

The goals of PA are to stop rainforest destruction and the loss of indigenous cultures and to find "workable ways in which the knowledge and wisdom inherent in both traditional cultures and the modern world can blend into a new global vision of sustainability for us all."

A brochure about PA is available on its website, which also includes an annual report, descriptions of rainforest trips, and membership information.

Rainforest Action Network (RAN)
221 Pine Street, Suite 500
San Francisco, CA 94104
Phone: 415.398.4404
Fax: 415.398.2732
http://www.ran.org/

In 1985, the Rainforest Action Network (RAN) was set up to work with other environmental organizations around the world on major campaigns to protect rainforests and the human rights of those living in and around those forests. The network convened the first international rainforest conference, which brought together activists from thirty-five organizations to develop action plans that helped catalyze the world movement to protect rainforests.

One of RAN's first action campaigns was a boycott of the Burger King restaurant chain to protest the company's practice of importing inexpensive beef from tropical countries where cattle ranching destroyed rainforests. Rainforest Action Groups, initially formed to conduct the campaign, have continued since then to work at the grassroots level to draw public attention to human activities that result in massive deforestation. RAN and the Action Groups have focused on the tropical timber trade, calling for a ban on the import and consumption of tropical timber products in the United States and other industrialized nations.

In 1997, RAN launched a campaign to end the market for old-growth redwood trees in the Pacific Northwest, later expanding the campaign to trade in all old-growth trees, including teak, Sitka spruce, rosewood, and greenheart. RAN has also conducted a home improvement campaign, using demonstrations, civil disobedience, celebrity pronouncements, and other actions to convince seven home improvement companies to help save old-growth forests and to commit themselves to reforming the global forest products industry. By the fall of 2000, companies pledging to eliminate the sales of wood from endangered forests included Home Depot, Lowe's Home Improvement Warehouse, Menard's Inc., Home Base Home Improvement Warehouse, Lanoga Corp., Wickes Lumber, and 84 Lumber Company.

This campaign and others are described on RAN's website, which also includes an educator's curriculum packet that can be downloaded, fact sheets for students and teachers, and membership information.

Rainforest Alliance (RA)
65 Bleeker Street
New York, NY 10012
Phone: 212.677.1900
Toll-Free: 888.MY EARTH
Fax: 212 941.4986
http://www.rainforest-alliance.org/

The primary mission of the Rainforest Alliance (RA) "is to develop and promote economically viable and socially desirable alternatives to tropical deforestation." To that end, RA helps with the development of forest products in cooperation with indigenous groups and other local organizations. Projects provide long-term, stable income for people living in or near tropical forests.

One major RA activity is the SmartWood Certification Program, which has been operating since 1989 and awards certificates to well-managed sources of tropical woods. Forest management operations that adhere to the program's environmental and social guidelines receive the SmartWood label to place on the wood that they sell. SmartWood-certified sources have met strict standards that ensure forests are managed in nondestructive ways and that local peoples receive long-term benefits (see SmartWood entry below). Other RA activities include promoting sustainable cultivation and harvesting of medicinal plants, conducting fund-raisers that benefit debt-for-nature swaps, supporting field research in tropical forests, and providing grants to community programs struggling to earn a livelihood without harming tropical forests or their wildlife.

The SmartWood program is explained in detail on the RA website, which also includes access to curriculum materials for educators, rainforest fact sheets and resource materials, a mission statement, and contact information.

Rainforest Information Centre (RIC)
Box 368
Lismore, NSW 2480
Australia
Phone: 61.2.66.218505, 213294
http://forests.org/ric/

A nonprofit volunteer organization in Australia, the Rainforest Information Centre (RIC) began after a successful struggle to save the subtropical rainforests of New South Wales, Australia, in the early 1980s. Since then, RIC has been involved in campaigns

and projects to protect rainforests and the legitimate development aspirations of rainforest peoples. Its website includes a children's page, information about rainforests, and descriptions of its
campaigns.

Save America's Forests (SAF)
4 Library Court SE
Washington, DC 20003
Phone: 202 544.9219
http://www.saveamericasforests.org/

Founded in 1990 as a nonprofit organization, Save America's
Forests (SAF) is a nationwide coalition of grassroots environmental groups, public interest organizations, and responsible businesses and individuals working to pass strong, comprehensive
federal laws to protect U.S. forest ecosystems. One purpose of the
organization is to bring grassroots activists to Washington, D.C.,
to testify before the U.S. Congress and to help educate members
of Congress about the principles of biodiversity and the need to
change the way national forests are managed. In 1997, the SAF
coalition began lobbying for passage of the Act to Save America's
Forests, which has been endorsed by 600 eminent scientists. If
passed, the federal law would permanently ban logging and road
building in core reserves of biological diversity in the ancient
forests of the Northwest and in more than 100 "special areas" that
are not otherwise protected. The law would ban clear-cutting and
tree farms on federal forest lands. Although the act has not yet
passed, at the end of 1999 President Clinton directed the U.S. Forest Service to devise plans to permanently protect 40 million acres
of national forest lands by banning road building.

Access to the text of the proposed law and other environmental and natural resources legislation are available on the SAF
website. In addition, the site explains how to join the organization and provides contact information.

Sierra Club
85 Second Street
San Francisco, CA 94105
Phone: 415.977.5500
Fax: 415.977.5799
http://www.sierraclub.org/

Naturalist John Muir founded the Sierra Club (SC) in 1892, and
since then the environmental organization has established chap-

ters throughout the United States to support the responsible use of the earth's ecosystems and resources. Through its various litigation and public education programs, the club has helped establish national parks and wilderness preserves.

These programs are described on the SC website, which also encourages grassroots support for a variety of programs to protect the environment, including efforts to reduce deforestation.

SmartWood
Goodwin-Baker Building
61 Millet Street
Richmond, VT 05477
Phone: 802.434.5491
Fax: 802.434.3116
http://www.smartwood.org/

A program of the Rainforest Alliance (see entry above), Smart-Wood was initiated in 1989 as the first forest management certification program of its kind. At first the program focused on tropical forests, but now it works for all forest types worldwide. Specialists at Rainforest Alliance offices in New York and Vermont collaborate with nonprofit organizations around the world to provide independent, objective evaluation of forest management practices, forest products, and timber sources and companies, enabling the public to identify products and practices that do not destroy forests. Through certification and use of the SmartWood label, the program provides a commercial incentive for forest managers to adopt sustainable forestry practices. SmartWood certifies forest products that come from "sustainable" or "well-managed" forests. Candidate sources may include a natural forest, a plantation, a large commercial operation, or a small-scale community project. SmartWood also certifies companies that process, manufacture, or sell products made from certified wood.

The SmartWood program also encourages the reuse of wood. A "SmartWood–Rediscovered Wood" seal of approval is awarded to products made from reclaimed wood. Such wood sources include demolished buildings, dead and fallen trees, unproductive orchard trees, usable and safe wood from landfills, and byproducts from secondary manufacturers. Details of the SmartWood programs are available on the SmartWood website.

Smithsonian Institution (SI)
Washington, DC 20560
Phone: 202.357.2627

Fax: 202.786.2377
http://www.si.edu/

Along with preserving artifacts and works of art that reflect America's cultural heritage, the Smithsonian Institution (SI), founded in 1846, sponsors scientific research and programs to protect the environment. Among its many programs, the SI established the Monitoring and Assessment of Biodiversity Program (SI/MAB) in 1986 to learn which species live in a specific forest, to determine how they interact with each other and with the environment around them, and to track those species over time. By conducting research, providing education and training courses, and monitoring forests around the world, SI/MAB has promoted biodiversity conservation in twenty-three countries. "SI/MAB works with governments, academic institutions, national and international organizations, industries and local communities to develop biodiversity monitoring and assessment programs. . . . In addition, SI/MAB has its own research sites in such places as Beni Biosphere Reserve in Bolivia, Luquillo Biosphere Reserve in Puerto Rico, Manu Biosphere Reserve in Peru, the U.S. Virgin Island Biosphere Reserve in St. John, Guatopo National Park in Venezuela, Kejimkujik National Park in Canada, and Hainan Island in China, among others."

SI/MAB projects are described on some of the numerous sites accessed from the SI homepage. The SI website also includes hundreds of educational resources on biodiversity and rainforests.

South American Explorers (SAE)
126 Indian Creek Road
Ithaca, NY 14840
Phone: 607.277.0488
Fax: 607.277.6122
http://www.samexplo.org/

The South American Explorers Club was founded in 1977, and in 1999 the organization changed its name to South American Explorers (SAE). This nonprofit, scientific, educational organization was founded primarily as an information center, advancing and supporting diverse forms of scientific field exploration and research in such areas as biology, geography, and anthropology. Along with the New York headquarters, SAE maintains offices in Lima, Peru, and Quito, Ecuador. It disperses contributions for scientific research and exploration and encourages information exchange among scientists and travelers from all nations who explore Central and South America. Another aim of the organization is to awaken greater interest in and appreciation for wilderness conservation and wildlife protection.

The SAE website posts information on Latin America, including trip planning and travel conditions, and presents news items about South American countries.

Survival International (SI)
11–15 Emerald Street
London WC1N 3QL
United Kingdom
Phone: 020.7242.1441
Fax: 020.7242.1771
http://www.survival.org.uk

Survival International (SI) is a nonprofit British-based group founded in 1969 to campaign for the rights of threatened tribal people, helping to ensure indigenous people's right to self-determination and control of natural resources. A worldwide organization, SI has supporters in eighty-two countries. It conducts letter-writing and education campaigns and embassy vigils, lobbies government officials, and works with local indigenous groups to protect tribal people's rights. Through its work in Brazil, for example, SI forced the government to recognize Yanomami land in 1992. According to SI, "Campaigns are not only directed at governments, but at companies, banks, extremist missionaries, guerrilla armies, museums, narrow minded conservationists and anyone else who violates tribal peoples' rights. Survival was the first organization to draw attention to the destructive effects of World Bank projects—now recognized as a major cause of suffering in many poor countries." SI's educational programs attempt "to demolish the myth that tribal peoples are relics, destined to perish through 'progress,' [and] promote respect for their cultures and explain the contemporary relevance of their way of life."

The SI online bookstore lists all SI publications, which can be ordered over the Web. The website also includes contact information for offices in the United Kingdom, France, Spain, and Italy.

Student Conservation Association (SCA)
P.O. Box 550
Charlestown, NH 03603
Phone: 603.543.1700

Fax: 603.543.1828
http://www.sca-inc.org/

"Changing lives through service to nature" is the motto of the Student Conservation Association (SCA), "America's largest and oldest provider of national and community conservation service opportunities, outdoor education, and career training for youth." SCA volunteers and interns work in national parks, forests, refuges, and urban areas in all 50 states. SCA's mission is to develop "conservation leaders and inspire lifelong stewardship of our environment and communities by engaging young people in hands-on service to the land."

On its website, SCA maintains one of the nation's largest conservation career employment databases, "Earth Work Online." The site also contains "Fast Facts" about SCA, describes volunteer and career opportunities in conservation, and tells users how to contact the organization.

Trees for the Future (TFTF)
P.O. Box 7027
Silver Spring, MD 20907.7027
Phone: 800.643.0001
Fax: 301.565.5012
http://www.treesftf.org/

Since it was founded in 1979, the nonprofit organization Trees for the Future (TFTF) has been establishing self-help projects worldwide that encourage protection of the world's remaining forests and bring life back to lands that have been destroyed by erosion, flooding, desertification, and infertility. TFTF concentrates on planting fast-growing trees in African, Asian, and Latin American villages. The trees can be harvested for fuelwood and building materials and help restore impoverished lands.

On its website, TFTF posts its position on issues of sustainable land management and provides access to "The Global Cooling Action Center," which lists ways to reduce the threat of global climate change.

Tropical Rainforest Coalition (TRC)
21730 Stevens Creek Blvd., Suite 102
Cupertino, CA 95014
http://www.rainforest.org/

Established in 1991, the Tropical Rainforest Coalition (TRC) is an all-volunteer, nonprofit organization that believes individuals can make a difference in preserving rainforests and the culture of indigenous people. The TRC's main effort is to build "coalitions between individuals, corporations, businesses, learning institutions and nongovernmental organizations seeking to reverse the unsustainable use of the rainforests." Through its "Save-An-Acre" program, the TRC purchases and manages rainforest sites and develops low-impact ecotourism programs. In addition, the coalition maintains a Tropical Bird Rescue program for mistreated and abandoned birds.

The TRC website includes descriptions of these programs and of the coalition's accomplishments, records of income and expenses, and information on how to contact and support TRC.

Union of Concerned Scientists (UCS)
2 Brattle Square
Cambridge, MA 02238
Phone: 617.547.5552
Fax: 617.864.9405
http://www.ucsusa.org/

An independent nonprofit alliance, the Union of Concerned Scientists (UCS) was founded in 1969 at the Massachusetts Institute of Technology by faculty members and students concerned about the misuse of science and technology in society. The UCS called for "the redirection of scientific research to pressing environmental and social problems." Today citizens and scientists across the United States are part of this advocacy group that supports rigorous scientific analysis in efforts to help "build a cleaner, healthier environment and a safer world." Scientists and engineers of the UCS work with their colleagues across the United States on studies focusing on such concerns as renewable energy options, the impacts of global warming, and biological diversity. Reports of UCS research are distributed to local, state, and federal government policymakers and the news media.

The UCS website includes descriptions of its programs, information on action campaigns, reports, and news releases.

United Nations Environment Programme (UNEP)
United Nations Avenue, Gigiri
P.O. Box 30552
Nairobi, Kenya
Phone: 254.2621234
Fax: 254.2.624489/90
http://www.unep.org/

The United Nations Environment Programme (UNEP) is one of a number of UN programs that help protect the global environment and manage natural resources. Established in 1972 as a result of the UN Conference on the Human Environment in Stockholm, Sweden, UNEP oversees the work of other UN agencies working in such areas as climate change, desertification control, water quality, and deforestation.

The UNEP website includes access to its annual report, *Global Environmental Outlook*, which can be downloaded, as well as legal documents on the environment, updates on global environmental events and awards, and information on the annual World Environment Day.

The Wilderness Society (TWS)

1615 M Street NW
Washington, DC 20036
Phone: 800.843.9453
http://www.wilderness.org/

Devoted to preserving wilderness and wildlife, The Wilderness Society (TWS) has established programs to protect America's prime forests, parks, rivers, and shorelands. Since its founding in 1935, the society has fostered an American land ethic and worked for federal legislation that sets aside wilderness areas for preservation. TWS has been instrumental in measures to help protect the Tongass National Forest in Alaska, and has joined efforts to conserve other wilderness areas throughout the United States.

The website of TWS includes access to fact sheets, historical information, position papers, and ways to support and contact the organization.

World Wildlife Fund for Nature (WWF)

Avenue du Mont-Blanc
CH-1196, Gland
Switzerland
Phone: 41.22.364.91.11
Fax: 41.22.364.53.58
http://www.panda.org/home.cfm

Founded in 1961, the World Wildlife Fund (WWF) became the World Wildlife Fund for Nature in 1986, although U.S. and Canadian organizations have retained the original name. WWF is the largest conservation organization working worldwide to protect endangered wildlife and wildlands. It is dedicated to "reversing the degradation of our planet's natural environment and to building a future in which human needs are met in harmony with the international WWF network," according to its website. A number of WWF conservation projects are under way in the tropical forests of Latin America, Asia, and Africa. For example, in the Brazilian state of Pará in eastern Amazonia, WWF is helping several communities find alternatives to the slash-and-burn pattern of agriculture. Local people are beginning to harvest valuable products and gain a fair return for their goods, which helps motivate preservation of the rainforest.

The WWF website includes information about where the organization has projects, descriptions of endangered species and spaces, forests, climate change, and other environmental issues.

World Forestry Center (WFC)
4033 SW Canyon Road
Portland, OR 97221
Phone: 503.228.1367
Fax: 503.228.4608
http://www.worldforestry.org/

The World Forestry Center (WFC) was founded in 1971 to educate the public about the importance of well-managed forests. The WFC has created special indoor and outdoor exhibits on forestry and forests of the world and at its demonstration tree farm/forest near Portland.

The WFC website describes its history, the museum it operates, and its education programs, and it lists resources for educators and provides contact information.

World Resources Institute (WRI)
10 G Street NE, Suite 800
Washington, DC 20002
Phone: 202.729.7600
Fax: 202.729.7610
http://www.igc.org/wri/

The World Resources Institute (WRI) was founded in 1982 and later merged with the Center for International Development and Environment. The WRI is a policy research center that assists governments, international organizations, and the private sector with basic policy decisions to balance human needs and economic growth with the preservation of natural resources. The WRI focuses on such issues as biological diversity, tropical forests, and incentives for sustainable development. In nonindustrialized nations, WRI provides technical assistance for government agencies and NGOs working to manage resources sustainably.

A major international effort of the WRI is Global Forest Watch, a program that "combines on-the-ground knowledge with digital and satellite technology to provide accurate forest information to anyone with access to the Internet." Global Forest Watch "identifies and promotes successful forest management practices, enables governments to better manage their forests, and provides local groups with the information they will need to participate in the management of their forests."

Global Forest Watch reports can be downloaded from the WRI website, which also includes facts and figures on such topics as biodiversity, climate change, and forests; a summary, data tables, and other information from the Global Resources 2000–2001 report; and environmental reports on regions around the world.

Worldwatch Institute (WI)
1776 Massachusetts Avenue NW
Washington, DC 20036
Phone: 202.452.1999
Fax: 202.296.7365
http://www.worldwatch.org/index.html

The well-known and respected Worldwatch Institute (WI) was founded in 1974 "to inform policymakers and the public about the complex links between the world economy and its environmental support systems." WI analyzes data from hundreds of research sources, including scientists and international organizations, and disseminates information through major media around the world and on its website.

Articles from WI's magazine, *World Watch,* can be ordered and downloaded from its website. In addition, the site contains a list of speakers on such topics as biodiversity and climate change, news items, and alerts to prompt action on environmental issues.

7

Selected Print and Nonprint Resources

Selected Print Resources

Since the 1960s, an increasing number of books have been published on rainforest topics, with a proliferation of titles appearing during the 1980s and early 1990s. Some of the earlier publications are now out of print, but usually they are available in public and university libraries. By 2000, the number of books on rainforests had increased even more, including revisions of earlier titles as well as dozens of new works on species extinction and biodiversity.

Many of the books listed in this chapter were written for a general readership; some are for readers who are looking for technical information and data on the world's rainforests; and others are for conservationists and environmental activists. Whatever their background, all readers who are interested in tropical or temperate rainforests should be able to find references suited to their particular purposes in the list that follows.

Atlases, directories, and guides that contain information on rainforests and ecotourism make up the first part of the listing. Following this section are alphabetical listings of books on a variety of subjects pertinent to the study of rainforests: books on biodiversity and species extinction; general works on the environment with some specific focus on rainforests or information related to rainforests, a few titles on global warming, with sections explaining how climate change is related to deforestation, works on indigenous people, books on sustainable society and the role of rainforest preservation in sustainability, and finally, books that focus on temperate and on tropical rainforests. Publishers' postal addresses and website URLs, if available, are listed at the end of this chapter.

Atlases, Directories, and Guides

Allen, John J. 1997. *The Student Atlas of Environmental Issues.* New York: WCB/McGraw-Hill. 112p.

This atlas is an educational tool for exploring the human impact on the air, waters, biosphere, and land in every major world region. A combination of maps and data helps students understand the dimensions of the Earth's environmental problems and the geographical basis of these problems.

Ashworth, William. 1991. *The Encyclopedia of Environmental Studies.* New York: Facts on File, 1991. 470p.

This reference work, although now somewhat dated, provides a comprehensive introduction to diverse environmental issues, including rainforest protection. The encyclopedia contains more than 3,000 entries defining environmental terms from many diverse disciplines, such as biology, botany, chemistry, economics, geography, and geology. Entries also describe individuals who have had an impact on the environmental movement, from Edward Abbey to James Watt. Major U.S. environmental laws, regulations, and regulatory agencies and significant events that led to environmental protection are explained. Diagrams, tables, and cross-references throughout enhance the reader's understanding of concepts and terms.

Boo, Elizabeth. 1990. *Ecotourism: The Potentials and Pitfalls.* 2 vols. Baltimore, MD: World Wildlife Fund. Vol. 1, 85p; vol. 2, 123p.

Among the early books on ecotourism, these two volumes focus on Latin America and the Caribbean, evaluating the economic and environmental impacts of ecotourism. Country case studies are included in the second volume.

Brown, Lester R., and the Worldwatch Institute. 2000. *State of the World: A Worldwatch Institute Report on Progress toward a Sustainable Society.* New York: W.W. Norton & Company. 250p.

Published since 1984, the annual *State of the World* report is issued in at least thirty languages and is the most widely used analysis of public policy in the world. This desktop guide presents the results of an "annual physical" of the world, including chapters relating to rainforest issues. Numerous figures and tables are included.

Burton, John, ed. 1998. *The Atlas of Endangered Species.* 2nd ed. New York: Macmillan. 256p.

Written by a team of experts, this atlas describes and illustrates numerous animal and plant species around the world in danger of extinction. With colorful illustrations and maps, the atlas is organized by regions and includes various focus sections, including one on tropical timber that describes endangered timber resources in Africa, the Philippines, and Latin America. There is a final section on "Conservation in Action," lists of endangered wildlife and of conservation organizations, and an extensive bibliography.

Emmons, Louise H., and Francois Feer. 1990. *Neotropical Rainforest Mammals: A Field Guide.* Chicago: University of Chicago Press. 550p.

Illustrated with beautiful photographs by Francois Feer, this field guide is a directory of mammals one might encounter in Central and South America at elevations below 1,000 meters. It contains information about approximately 500 species, including measurements and markings, geographic variation, natural history, and geographic range.

Gordon, Rue E., and Martha Riecks-Tracey, eds. 2000. *2001 Conservation Directory: A Guide to Worldwide Environmental Organizations.* New York: Lyons Press. 544p.

Revised each year, this National Wildlife Federation directory lists governmental and nongovernmental organizations engaged in conservation efforts at the state, national, or international level. Entries include the titles and names of personnel along with brief descriptions of agencies and organizations, educational institutions, databases, federally protected conservation areas, and much more information useful to environmental activists, students, outdoor writers, science editors, natural-resource agencies, and other individuals interested in wildlife and ecology.

Holing, Dwight. 1991. *earthTrips: A Guide to Nature Travel on a Fragile Planet.* Venice, CA: Living Planet Press. 209p.

A Conservation International book, this is one of the early ecotourism guides that describes an increasingly popular way to travel. In the author's words, ecotourism is "ecologically sensitive travel that combines the pleasures of discovering and understanding spectacular flora and fauna with an opportunity to contribute to their protection." The book explains the benefits of such travel: By attracting tourists to national parks and other reserves, many of which are rainforests, local people are able to earn income and at the same time protect natural areas. Another purpose of ecotourism is to promote environmental awareness and conservation worldwide. This book explains how to "travel with a cause" in seven regions of the world.

Mason, Robert J., and Mark T. Mattson (contributor). 1990. *Atlas of United States Environmental Issues.* New York: Macmillan. 252p.

Described as a research and teaching tool, this atlas can be a primary source for students and others investigating environmental issues in the United States. It includes four-color maps, charts, graphs, and diagrams. Text covers such topics as agricultural lands; forests and forestry; parks, recreation, and wildlife; the environment and politics; and major environmental legislation.

Stein, Edith C. 1992. *The Environmental Sourcebook.* New York: Lyons & Burford. 264p.

Written by a physician specializing in environmental medicine and president of the Environmental Data Research Institute, this sourcebook is designed to help citizens understand major environmental issues, such as rainforest destruction, endangered species, and atmospheric pollution. The book also identifies foundations that fund environmental causes and explains where readers can find further information on particular ecological issues.

Sunkin, Maurice, David M. Ong, and Gold Phil. 1998. *Environmental Law.* London: Cavendish Publishing. 798p.

Students taking environmental law courses should find this book helpful. It discusses the principal sources of environmental law and includes primary source materials from international, European and British law, including the main treaties and other international documents, case law, European instruments, and British legislation.

Trzyna, Thaddeus C., ed. 1996. *World Directory of Environmental Organizations.* 5th ed. Claremont, CA: California Institute of Public Affairs. 167p.

This is a handbook of national and international organizations and the programs of governments and nongovernmental organizations concerned with the protection of Earth's resources.

Biodiversity and Species Extinction

Bates, Marston. 1988. *The Forest and the Sea.* New York: Lyons & Burford. 288p.

This is a classic work on the general principles of the organization of the biological community, which examines the patterns of relationships between individuals, populations, and species. A naturalist, Marston Bates eloquently describes the different biomes: the seas, the rivers, the rainforests, the woodlands, and the deserts. He places humans as a natural part of this system rather than as a special phenomenon.

Daily, Gretchen C., ed. *Nature's Services: Societal Dependence on Natural Ecosystems.* Washington, DC: Island Press. 416p.

Several dozen authors have contributed to this book, which includes chapters that define varied ecosystem services, such as the interaction of climate and life, biodiversity, and the ecosystem services of soil. "Natural Pest Control Services and Agriculture," "Freshwater Ecosystem Services, and "The World's Forests and Their Ecosystem Services" are among the twenty chapters.

Ehrlich, Paul R., and Anne H. Ehrlich. 1981. *Extinction—The Causes and Consequences of the Disappearance of Species.* New York: Random House. 294p.

Still a relevant text, the Ehrlichs' book shows with strong, reasoned arguments and documentation how the extinction of only a single species could lead to disaster. They explain how humans have endangered certain species, how they benefit from endangered species, and what can be done to protect these species.

Huxley, Anthony. 1992. *Green Inheritance—The World Wildlife Fund Book of Plants.* New York: Four Walls Eight Windows Press.

This book is a revision of the 1985 edition published by Anchor/ Doubleday. Although not specifically about rainforests, it describes plants as the basis of the food cycle, explaining that as the first to disappear, plants signal the next step: the disappearance of animals, including humans. Many chapters point out the importance of rainforest plants and their products. Individual chapters include discussions of plants as sustenance, plants as medicine, plants as crops, and plants as objects of beauty. There is also a brief section explaining how the loss of vegetation contributes to the greenhouse effect. Color photographs and drawings are included.

Kaufman, Les, and Kenneth Mallory, eds. 1993. *The Last Extinction.* 2nd ed. Cambridge, MA, and London: The MIT Press. 208p.

Based on a public lecture series called "Extinction: Saving the Sinking Arc" held in Boston, this book focuses on ecology and species extinction. The collection of essays, written by various experts, is designed "to awaken the general public to the issues underlying the notion that we must prevent a 'last extinction,' . . ." arguing that even though "mass extinction is in progress . . . it can be postponed indefinitely." The first chapter is an overview of the complex problem of species extinction, and the second reviews evidence of past extinctions. Ongoing destruction of tropical forests is the subject of chapter three, while the fourth essay describes "Vanishing Species in Our Own Backyard." A fifth essay explains the need to transform zoos and aquaria into refuges for endangered species. Finally, "Life in the Next Millennium—Who Will Be Left in Earth's Community?" argues for stewardship of the Earth and suggests ways to preserve it. Graphs, charts, and black-and-white photographs are included.

Kennedy, Michael. 1990. *Australia's Endangered Species.* Upper Saddle River, NJ: Prentice Hall Press. 192p.

A variety of conservation experts have contributed to this now out-of-print work, which opens with an essay by Norman Myers, internationally known British consultant and author of many books on the environment and development. Chapters cover Australia's threatened mammals, birds, reptiles, amphibians, and fish. The book is illustrated with color photographs by Australia's top wildlife photographers. There are also extensive lists of Australian plant and animal species that are threatened with extinction or presumed extinct.

Levin, Simon A. 2000. *Encyclopedia of Biodiversity.* San Diego, CA: Academic Press. 4700p.

This five-volume encyclopedia, which includes numerous articles by scholarly contributors, focuses on the critical importance of biodiversity, examines the services that biodiversity provides, and describes measures to protect it. Major themes of the work include the evolution of biodiversity, systems for classifying and defining biodiversity, ecological patterns and theories of biodiversity, and an assessment of contemporary patterns and trends in biodiversity. Sir Robert May, chief scientific adviser to the British government, notes: "My participation in the development of this Encyclopedia has been a very encouraging experience, particularly in providing an opportunity to see how broadly interest in biodiversity has spread. The Encyclopedia has deliberately been organised in a most inclusive way, and this will help people of all backgrounds learn more about their world and the other living things they share it with." Scholars, activists, and decision makers should find this a helpful resource. It includes approximately 1,000 figures and tables and more than 3,000 glossary entries.

Oates, John F. 1999. *Myth and Reality in the Rain Forest: How Conservation Strategies Are Failing in West Africa.* Berkeley and Los Angeles: University of California Press. 310p.

In this book, John Oates disagrees with the theory that wildlife can best be protected through the promotion of human economic development. He bases his work on his extensive experience as a primate ecologist working on rainforest conservation projects in Africa and India. Oates contends that linking conservation to economic development has had disastrous consequences for many wildlife populations, especially in West Africa. In his view, the well-being of the very poor is more likely to be promoted by large-scale political, social, and economic reforms than by community development schemes associated with conservation projects. He also argues that sustainable development is failing to protect species and parks, and that nature should be protected for its own sake.

Schultes, Richard Evans, and Robert F. Raffauf. 1990. *The Healing Forest: Medicinal and Toxic Plants of the Northwest Amazonia.* Portland, OR: Dioscorides Press. 484p.

An ethnobiologist and a phytochemist are the authors of this work, which describes 1,516 plant species in the Northwest Amazonian rainforest. Little or no chemical analysis has been conducted on many of these species, and experts estimate that there could be as many as 80,000 plants yet to be discovered and researched in the area. The authors call attention to the need for conservation and emphasize the importance of ethnobiology. It is their hope that this reference will alert readers to the importance of Amazonian plants and will also encourage the further study of indigenous folklore.

Smith, Charles H. 2000. *Biodiversity Studies: A Bibliographic Review.* Lanham, MD: Scarecrow Press. 480p.

This work, which focuses on the period 1986 to 1998, includes two bibliographies with brief annotations. One bibliography lists monographs and short articles, and the other is a selection of serial publications of special issues dealing with biodiversity. Other closely related topics are included in the entries, such as environmental ethics and law, sustainable development, forestry, and climatology. The publication, compiled by a science librarian, would be of interest to scientists and educators, as well as some undergraduate and high school students.

Stearns, Beverly Peterson, and Stephen C. Stearns. 1999. *Watching, from the Edge of Extinction.* Yale University Press. 288p.

Ten case studies by Beverly Stearns and her husband dramatically show how scientists and environmentalists fight to keep endangered species alive. The authors blame much of the danger to species survival on politics, greed, and corruption.

Stein, Bruce A., Lynn S. Kuner, and Jonathan S. Adams, eds. for The Nature Conservancy and the Association for Biodiversity Information. 2000. *Precious Heritage: The Status of Biodiversity in the United States.* New York: Oxford University Press. 416p.

This illustrated book, which includes helpful maps and photographs, describes the biological diversity and wealth of America's lush forests and an array of other ecosystems. Information collected over twenty-five years by a network of natural heritage programs not only documents U.S. biodiversity, but also shows what and how species and ecosystems are threatened and what is needed to protect them.

Vergoth, Karin, and Christopher Lampton. 2000. *Endangered Species.* (Revised edition. Danbury, CT: Franklin Watts/Grolier. 112p.

A young-adult title, this book examines Earth's natural resources, including habitats that have been destroyed, and explains what people need to do to preserve them.

Wilson, Edward O. 1992. *The Diversity of Life.* Cambridge, MA: Harvard University Press. 464p.

Called a "bible of the biosphere," this book guides the reader through evolutionary time, describing and showing with color photographs and illustrations the diversity of the biological world. Wilson, who is considered the dean of biodiversity studies, not only describes the threats that human activities pose to our biological wealth but also shows how the preservation of that wealth can be a part of economic development.

General Environmental Subjects

Abramovitz, Janet N., Lester R. Brown, Seth Dunn, Christopher Flavin, Hilary F. French, Gary Gardner, Brian Halweil, Nicholas Lenssen, Lisa Mastny, Ashley T. Mattoon, Anne Platt McGinn, Sara Porter, Michael Renner, David M. Roodman, Payal Sampat, Michael Scholand, and Molly O. Sheehan. 2000. *Vital Signs 2000: The Environmental Trends That Are Shaping Our Future.* Washington, DC: Worldwatch Institute. 192p.

Written by the staff of the Worldwatch Institute, this is the ninth volume in a series that shows key global trends indicating environmental, economic, and social progress or regression. The data used as a basis for the "vital signs" are gleaned from documents gathered from scientists, international organizations, government, and industry.

Arnold, Ron, and Alan Gottlieb. 1998. *Trashing the Economy: How Runaway Environmentalism Is Wrecking America.* 2nd ed. Bellevue, WA: Merril Press. 670p.

Written by the leaders of the wise use movement, this book presents profiles of the top sixty environmental groups in America, showing where they get their money and what they do with it. The authors contend that environmentalists are destroying the natural resource base of the economy.

Bailey, Ron, ed. 1999. *Earth Report 2000.* New York: McGraw Hill. 362p.

This guide, prepared with the support of the Competitive Enterprise Institute, was edited by Ron Bailey, a writer for *Forbes* magazine and a PBS producer. The book's ten chapters, written by environmental researchers, evaluate the ecological condition of Earth and cover such topics as global warming and biodiversity. Forty-two illustrations are included.

Beckerman, Wilferd. 1996. *Through Green-Colored Glasses: Environmentalism Reconsidered.* Washington, DC: Cato Institute. 230p.

This book, written by an emeritus fellow of Balliol College, Oxford, and a former member of the Royal Commission on Environmental Pollution, presents a view contrary to that of most environmentalists and environmental scientists regarding the threat of natural resource depletion and global warming. In Beckerman's view, developing countries have low levels of economic development because they have not been able to produce or purchase the technology needed to protect the environment. His conclusion: "The best—and no doubt the only—route by which these countries can overcome their appalling environmental problems is to become richer."

Brown, Lester R., Christopher Flavin, Hilary French, Janet Abramovitz, Seth Dunn, Bary Gardner, Lisa Mastry, Ashley Mattoon, David Rodman, Payal Sampat, Molly O. Sheehan, and Linda Starke, ed. *State of the World 2001.* New York: W.W. Norton & Company. 256p.

Charts and graphs are an important part of this annual publication written by the staff of Worldwatch Institute. An important reference on the world's environment, it provides national leaders and concerned citizens with a comprehensive framework for the global debate about our future in the new century. This annual survey by the award-winning Worldwatch Institute has become an invaluable analysis of negative environmental trends and a guide to emerging solutions.

Easterbrook, Gregg. 1996. *A Moment on the Earth: The Coming Age of Environmental Optimism.* Reprint edition. New York: Penguin.

In this controversial book, Gregg Easterbrook argues that environmental conditions are improving around the globe and explains why he believes pollution is almost a problem of the past in the Western world. Predicting a "green" future, he calls for a new and more realistic view of ecology issues. He contends that technology will help prevent environmental catastrophes and that species extinction is a natural process.

Ehrlich, Paul R., and Anne H. Ehrlich. 1991. *Healing the Planet.* New York: Addison-Wesley. 336p.

In this work, these two well-known scientists focus on the causes—the underlying problems—rather than the symptoms of environmental destruction. The authors recognize that the disparity between the rich and poor nations must be addressed and that the problems of the Southern Hemisphere need to be resolved in order to make progress in terms of environmental protection. The book includes extensive source notes.

————. 1998. *Betrayal of Science and Reason: How Anti-Environment Rhetoric Threatens Our Future.* Revised edition. Washington, DC: Island Press. 335p.

Stanford University professors Paul and Anne Ehrlich have written an angry book to counteract what they call a "brownlash" by anti-environmentalists. In the Ehrlich's' view, politically motivated, anti-environmental groups have used pseudoscience and misinformation to discredit pressing environmental concerns such as ozone depletion, global warming, loss of biodiversity, and overpopulation, thus diminishing the seriousness of these issues in the public arena. As they have for more than two decades, the Ehrlichs discuss the damage to the environment caused by industrial emissions, toxic waste, and other pollutants. Backed by scientific consensus, they argue that the planet is in real peril, and they detail the ""inadequacies of "feel-good" books by such "brownlashers" as Gregg Easterbrook, Stephen Mark Plummer, and politician Dixie Lee Ray.

Goodstein, Eban. 1999. *The Trade-Off Myth: Fact and Fiction about Jobs and the Environment.* Washington, DC: Island Press. 195p.

Economist Eban Goodstein provides an in-depth examination of the deep-seated but ultimately mistaken American belief in a widespread jobs-environment trade-off. The book offers a readable and accessible analysis of the labor impacts of environmental regulation. Goldstein considers the roots and staying power of misperceptions regarding job security and environmental regulation, analyzes various models used to predict employment impacts and explains how changes in assumptions can drastically change predicted outcomes, lists and debunks, myth by myth, widely held perceptions about the impacts of environmental regulation on jobs, examines localized hardships caused by environmental protection measures within specific industries and regions and considers what can be done to mitigate those impacts, explores the revisionist view that environmental protection measures can actually create jobs, and looks at the jobs-environment issues that are likely to emerge as efforts are made to combat global warming.

Gore, Albert. 1992. *Earth in the Balance: Ecology and the Human Spirit.* New York: Houghton Mifflin. 407p.

As a U.S. Senator from Tennessee and as vice president of the United States, Al Gore has long worked for environmental protection. He explains in this book how he reached the conclusion that human civilization's "ravenous appetite for resources" threatens the ecological balance on Earth. Gore argues that we all need to reshape our attitudes and ideas about our relationship with nature if we are to save the planet from ecological disaster. He then describes a "Global Marshall Plan" with five strategic goals: stabilizing world population, developing environmentally appropriate technologies, changing the way people measure the economic impact of their decisions to include long-term costs to the environment, negotiating international agreements to protect the world's ecology, and educating people around the world on the need to conserve the Earth's resources. Gore also points out the need for social and political justice so that sustainable societies can exist. Extensive chapter notes and bibliography are included.

Lovelock, James. 1988. *The Ages of Gaia: A Biography of Our Living Earth.* New York: W.W. Norton & Company. 252p.

James Lovelock first spelled out his Gaia theory, named for the Greek goddess of Earth, in a 1979 best-seller, *Gaia: A New Look at Life on Earth.* With scientific data from several disciplines, *The Ages of Gaia* supports the theory that Earth is a living whole. Lovelock shows how human activities such as fossil fuel burning and deforestation that contribute to the greenhouse effect threaten the health of the planet and ultimately human life itself.

———. 1991. *Healing Gaia: Practical Medicine for the Planet.* New York: Harmony Books. 192p.

Building on his Gaia theory, Lovelock takes on the role of the "planetary physician" to diagnose and suggest cures for various environmental ailments. However, much of this book attacks the many critics of the controversial Gaia hypothesis—scientists who do not accept the concept that the Earth should be seen as one living organism.

McKibben, Bill. 1989. *The End of Nature.* New York: Random House. 226p.

"Nature, we believe, takes forever." So begins Bill McKibben's book, which shows that because of human activities, "our reassuring sense of a timeless future . . . is a delusion." McKibben argues that global warming and the depletion of the ozone layer are already destroying nature as we know it, and he makes a compelling plea that people act as caretakers of the Earth and custodians of all life-forms.

World Resources Institute in collaboration with the United Nations Development Programme, the United Nations Environment Programme, and the World Bank. 2000. *World Resources 2000–2001: People and Ecosystems—The Fraying Web of Life.* Washington, DC: World Resources Institute. 400p.

The condition of the Earth's five critical ecosystems and how to protect them is the focus of this volume, the ninth annual *World Resources* publication. Those five ecosystems—croplands, forests, coastal zones, freshwater systems, and grasslands—all help sustain human life. This report details how good management and current information about the capacity of ecosystems are important ingredients in the effort to preserve ecosystems so that they can continue to provide goods and services. Case studies and examples show how people worldwide are improving the way they manage ecosystems, and as in previous editions, the book also presents an overview of current environmental trends around the world.

Global Warming and Climate Change

Dale, Virginia, ed. 1993. *Effects of Land-Use Change on Atmospheric CO_2 Concentrations: South and Southeast Asia as a Case Study.* New York: Springer Verlag. 384p.

Authors contributing to this book describe the impact on the atmosphere of the changing use of land—notably deforestation—in the tropical forests of Asia. Approaching this problem from a variety of perspectives, the work uses historical data, socioeconomic analyses, computer modeling, and ecological field research. Although the research emphasizes atmospheric carbon dioxide and focuses on southeast Asia as a case study, the book's implications are global in their significance.

Harris, Paul G., ed. 2000. *Climate Change and American Foreign Policy.* Hampshire, UK: Palgrave. 302p.

This is a compilation of articles from varied contributors with different perspectives. The book introduces the issue of climate change in the context of U.S. foreign policy and its actions in climate change negotiations. The work covers the international negotiations leading to the Framework Convention on Climate

Change and the Kyoto Protocol, looking at the role of international norms in shaping U.S. climate change policy. Geared toward policy makers and environmental activists, the volume includes chapters on the politics of U.S. climate change policy, domestic influences on global environmental policy, and regulation theory.

Houghton, John T. 1997. *Global Warming: The Complete Briefing.* 2d ed. New York: Cambridge University Press. 240p.

Sir John Houghton, a world-renowned expert on climate change, explores the scientific basis of global warming in this comprehensive guide. The book discusses the action that governments, industry, and individuals can take to help lessen global warming. Although the text is appropriate for classroom use, it is also a helpful resource for the layperson who has some general scientific background.

Michaels, Patrick J., and Robert C. Balling, Jr. 2000. *The Satanic Gases: Clearing the Air about Global Warming.* Washington, DC: Cato Institute. 224p.

The authors of this work have often argued that global warming is fundamentally a mythical problem. Here again, they contend that there will be only a modest warming of the Earth over the twenty-first century, basing their findings on a variety of scientific studies cited.

Oppenheimer, Michael, and Robert H. Boyle. 1990. *Dead Heat: The Race Against the Greenhouse Effect.* New York: Basic Books. 268p.

"Humanity is hurtling toward a precipice. Left unchecked, the emissions of various gases, particularly carbon dioxide from fossil-fuel combustion and deforestation, are likely to alter the Earth's climate so rapidly and so thoroughly as to destroy much of the natural world. . . . But such an outcome is not inevitable." So begins the prologue to this book, written by a senior scientist with the Environmental Defense Fund and a senior writer for *Sports Illustrated.* The opening statements signal the content of this book: an action plan for ways that industries and individuals can reverse the trend toward global warming. Copious source notes are included.

Revkin, Andrew. 1992. *Global Warming: Understanding the Forecast.* New York: Abbeville Press. 180p.

Illustrated with dramatic color and black-and-white photographs and a variety of drawings, this book was published to
complement a major traveling exhibition by the same title prepared in conjunction with the American Museum of Natural History in New York City. Award-winning journalist Andrew
Revkin explains how scientists study past and present climate in
order to predict future changes and shows how human activities,
such as the burning of fossil fuels and deforestation, lead to a
buildup of carbon dioxide and other greenhouse gases. In a chapter entitled "Business as Usual," Revkin outlines the dire consequences of not taking action now to reduce CO_2 emissions. The
final chapter, "A Greenhouse Diet," describes what individuals
can do to reduce the risk of global warming. Included also are
lists of organizations, resources, and suggested reading.

Indigenous Peoples

Burger, Julian. 1990. *The Gaia Atlas of First Peoples: A Future for the
Indigenous World*. New York: Anchor/Doubleday. 191p.

In a clear and graphic style, this atlas is a unique source for anyone interested in the lives of first peoples, including indigenous
people of the rainforests. Fifty concise essays, illustrated with
photographs, maps, charts, and other graphics, describe Native
peoples worldwide and show how resource management, herbal
medicine, cooperation, and conflict resolution are intrinsic to
many indigenous cultures.

Clay, Jason W. 1988. *Indigenous Peoples and Tropical Forests: Models
of Land Use and Management from Latin America*. Cambridge, MA:
Cultural Survival,1988. 116p.

How do indigenous people of the tropical rainforests use and
sustain the resources of their regions? That is the basic question
explored in this report by the former director of Cultural Survival. The report summarizes activities such as hunting and gath-
ering and various types of agriculture that sustain indigenous
populations and their environment. A bibliography of more than
400 works is included.

Denslow, Julie Sloan, and Christine Padoch. 1988. *People of the
Tropical Rainforest*. Berkeley and Los Angeles: University of Cali-
fornia Press; Washington, DC: Smithsonian Institution Traveling
Exhibition Service. 231p.

In the words of this book's authors, "The people of the tropical rainforests of the world, while all one species with similar needs, capabilities, and tolerances, are also an enormously varied lot." In a collection of essays, contributors to this work describe the cultures of such groups as the Kayapo, the Lacandon Maya, the Pygmies of the Congo basin, and the Hmong and Lua of Thailand, and more recent arrivals to the rainforest who depend on slash-and-burn agriculture for their staple crops and on the forest and rivers for game and fish. Illustrated with color photographs, this book, as the preface notes, "offers no solutions to the 'rain forest problem'." But contributors do explore the diverse ways that people use the forest, describe the widespread impact of such business ventures as logging, plantation forestry, mining, and cattle ranching, and introduce some projections about the future of tropical rainforests and the people who live in them.

Good, Kenneth, and David Chanoff. 1991. *Into the Heart: One Man's Pursuit of Love and Knowledge among the Yanomama.* New York: Simon & Schuster. 352p.

In 1975 Kenneth Good, an anthropology student, began working with the Yanomama Indians of the Amazon rainforest. He lived in Yanomama communities for twelve years, studying their way of life, learning their language, and eventually marrying a woman chosen for him by the tribe. Although Good describes the violence the Indians sometimes inflict upon one another, he also shows the respect and kindness that prevails. This book is not only a story of the culture of an indigenous people but also a personal account of Good's relationship with his wife, Yarima, whom the headman selected for him when she was nine years old. Good waited for Yarima to come of age for marriage and fell in love with her. Because he was an outsider, he encountered problems with Venezuelan officials and nearly lost his wife, but he was able to take her with him when he returned to the United States to teach at Jersey City State College in New Jersey. The book is illustrated with photographs.

Hansen, Eric. 1988. *Stranger in the Forest: On Foot across Borneo.* Boston: Houghton Mifflin. 286p.

In 1982, Eric Hansen began his 1,500-mile walk through the Borneo rainforest, keeping copious notes of his adventures along the way. This book is not only a story of Hansen's journey but also a look at the Penan, indigenous people whose way of life is threatened modern-day development. A gifted storyteller, Hansen brings the rainforest to life with humor, compassion, and attention to detail.

Ricciardi, Mirella. 1991. *Vanishing Amazon.* New York: Abrams. 240p.

A renowned photographer, Mirella Ricciardi presents a photographic diary of her trip to the Amazonian region of western Brazil, including her observations of three Amazon Indian groups. In this work, which includes 220 photographs, she shows the daily life and culture of the Kempa, the Marubo, and the Yanomami and explains how Amazon Indian groups have joined together to fight for the preservation of their forest homelands. The text is brief, but anthropologist Marcus Colchester provides background information on each of the tribes.

Ritchie, Mark Andrew. 2000. *Spirit of the Rainforest: A Yanomamo Shaman's Story.* 2nd ed. Island Lake, IL: Island Lake Press. 288p.

The author of this adventure-filled and well-documented book allows a Yanomamo shaman to tell the story, lending authenticity. It makes the spirit world appear real and examines some of the stereotypes of and misconceptions about the indigenous people in Venezuela.

Sustainable Society

Brown, Lester R., Christopher Flavin, and Sandra Postel. 1991. *Saving the Planet: How to Shape an Environmentally Sustainable Global Economy.* New York: W.W. Norton & Company. 224p.

The first in a World Watch Institute series on the environment, this volume strives to answer the most fundamental question of today's generation: How can we create a world economy and not destroy the Earth in the process? The goal of saving the Earth evokes a vision of a global economy that does not compromise the prospects of future generations. To achieve this vision, the authors offer a way to restructure energy production systems, economic systems, and aid distribution programs. Copious chapter notes are included.

Costanza, Robert, ed. 1991. *Ecological Economics: The Science and Management of Sustainability.* New York: Columbia University Press. 525p.

Forty-two experts contributed to this technical work for college students. The volume, which is the result of the first biannual workshop of the International Society for Ecological Economics, links economics and ecology and shows how countries can develop economic policies that do not destroy global ecosystems. It is divided into three parts: "Developing an Ecological Economic World View, "Accounting, Modeling, and Analysis," and "Institutional Changes and Case Studies." Contributors explore the research, training programs, and complex techniques needed to deal effectively with environmental problems.

Darmstadter, Joel, ed. 1992. *Global Development and the Environment: Perspectives on Sustainability.* Washington, DC: Resources for the Future. 91p.

Written for those who have some understanding of economics, ecology, and sustainability, this book is an interdisciplinary compilation of essays addressing such problems as biodiversity, population control, the wise use of natural resources, and alternative energy sources. The volume provides an overview of sustainable development and includes practical solutions to environmental problems, along with actions that students and others can take to bring about sustainability.

Goodland, Robert, ed. 1990. *Race to Save the Tropics: Ecology and Economics for a Sustainable Future.* Washington, DC: Island Press. 219p.

The authors of the varied chapters in this book are what the editor calls "muddy-footed practitioners" who are committed to the concept that ecologically sound development makes economic sense and can become a reality. Along with sound scientific data, the authors provide practical examples of how economic ecology can be applied to forestry.

Hall, Anthony, ed. 2000. *Amazonia at the Crossroads: The Challenge of Sustainable Development.* London: University of London, Institute of Latin American Studies. 280p.

This book is based on the 1998 conference "Amazonia 2000: Development, Environment, and Geopolitics" held in London and organized by the Institute of Latin American Studies. The work examines past patterns of destructive resource extraction in Amazonia and shows how more acute awareness of the need to encourage more sustainable policies and practices has emerged at all levels, national and international. These practices include new production strategies, such as agroforestry, and innovative resource governance models, such as inland fisheries comanagement and agroecological zoning. How to integrate such policies and practices into mainstream development within Amazonia is the major challenge.

Harrison, Paul. 1992. *The Third Revolution: Environment, Population, and a Sustainable World.* New York: St. Martin's Press. 351p.

In this call for a third revolution (going beyond the agricultural and industrial revolutions) to achieve sustainable development, Paul Harrison explains how overconsumption, rapidly increasing population, and destructive technologies have contributed to global ecological problems. He describes disasters brought on in part by overpopulation, basing his discussion on his firsthand knowledge and fieldwork in Asian countries such as Bangladesh. Building on work by Paul Ehrlich, Harrison explains the equation "Population x Consumption x Technology = Environmental Impact" and shows with specific examples how population growth and environmental deterioration are linked. The book also includes specific steps that can be taken to stabilize population, such as granting women rights equal to those of men and providing access to birth control and health care.

Piasecki, Bruce, and Peter Asmus. 1990. *In Search of Environmental Excellence.* New York Simon & Schuster. 200p.

This book presents a variety of suggestions for ordinary citizens, public officials, policy makers, activists, and industry leaders who want to work toward resolving environmental problems, and also includes ideas for controlling deforestation. The book is essentially a blueprint for survival and a call to action to save the planet.

Roddick, Anita. 1991. *Body and Soul: Profits with Principles—The Amazing Success Story of Anita Roddick and the Body Shop.* New York: Crown Publishing. 256p.

This story of an environmental activist and business woman explains how Anita Roddick made millions for her investors without selling her soul. Roddick, who has become one of the most outspoken activist business women in the world, founded The Body Shop, manufacturing cosmetics and skin care products from "natural" sources and competing in what she calls the "nastiest industry in the world." Roddick describes how her company espouses social and environmental responsibility and acts upon those principles through "trade-not-aid" programs in the Amazon and other parts of the world.

Stone, Roger D. 1992. *The Nature of Development.* New York: Knopf. 304p.

Using evidence from various environmental organizations and development agencies, Roger Stone, of the World Wildlife Fund, shows how many international development efforts have failed to reduce the poverty of millions of people. As a result, the poor in developing nations have placed demands on global resources—for example, cutting trees on their land for meager profits or overfishing for food. Stone also describes some successful grassroots initiatives to conserve resources in developing countries.

Temperate Rainforests

Allaby, Michael. 1999. *Temperate Forests.* New York: Facts on File. 224p.

Part of a four-volume ecosystem series, Michael Allaby's book explores the ecology, biology, chemistry, history, and economics of the forest. Details on each ecosystem include its history, biology, wildlife, beauty, problems, and influence on culture. In addition, the work covers the role of the temperate forest in the modern economy; the evolution of forests and forestry; the portrayal of the forest in fairytales, legends, and Shakespeare; and future prospects for sustainable forest management. Full-color photographs, line drawings, and tables are included.

Davis, John, ed. 1991. *The Earth First! Reader: Ten Years of Radical Environmental Journalism.* Layton, UT: Gibbs Smith Publishers. 240p.

Cofounder of the Earth First! movement, Dave Foreman, wrote the foreword for this compilation of forty feature articles from the discontinued *Earth First! Journal.* The articles provide a history of this radical environmental movement and the concepts behind it. Using confrontational tactics and on occasion, illegal methods, the Earth First! movement gained widespread publicity during the 1980s, when it first called attention to environmental destruction, particularly in the ancient forests of the Pacific

Northwest. The zeal and passion of the writers are reflected in the articles, but at times they seem to accuse almost everyone of raping the Earth. A final section explains why the movement split in 1990.

Devall, Bill, ed. 1995. *Clearcut: The Tragedy of Industrial Forestry.* San Francisco, CA: Sierra Club Books. 304p.

The loss of North American forests is the focus of this book, which calls for a reversal of the needless devastation of these resources and shows how the natural order can be restored. Contributors include fifteen leading ecologists and activists, among them Earth First! cofounder Dave Foreman; Jim Cooperman, director of the British Columbia Environmental Network; entrepreneur-environmentalist Doug Tompkins; conservation biologist Reed Noss; and some of America's premier nature photographers, among them Galen Rowell, Jenny Hager, and David Hiser. The book contains ninety-eight color photographs, seventy-eight black-and-white photographs, and maps.

Dietrich, William. 1992. *The Final Forest: The Battle for the Last Great Trees of the Pacific Northwest.* New York: Simon & Schuster. 290p.

A Pulitzer Prize–winning correspondent for the *Seattle Times* is the author of this anecdotal and clearly presented account of the people affected by controversies over the ancient forests of the Pacific Northwest, particularly the Olympic National Forest. In each chapter, Dietrich presents stories of people from a broad range of backgrounds—ecologists, mill workers, biologists, industrialists, truckers, politicians, forest rangers, and others. Yet the book also shows that the forest "is perhaps a metaphor for all that we cherish and exploit on earth." In his concluding words, Dietrich makes his point: "This book is, of course, written on a tree. . . . I hope you read it for whatever understanding it provides. Then, when you get a chance, go and read the living things that it came from."

Durbin, Kathie. 1999. *Tongass: Pulp Politics and the Fight for the Alaskan Rainforest.* Corvallis, OR: Oregon State University Press. 336p.

A former environmental journalist for the Portland *Oregonian,* Kathie Durbin tells the story of the Tongass National Forest, the world's largest temperate rainforest. The book shows how the

U.S. government and federal legislation passed in the 1950s allowed two paper pulp companies to establish operations in and clear-cut significant portions of the Tongass. The book details how paper mills brought jobs and growth to a sparsely settled region but also brought ecological destruction. The book also shows how the timber industry "broke labor unions, drove competitors out of business, and controlled politicians and the U.S. Forest Service." Durbin tells stories about those most affected, including Tlingit Indians who saw their traditional hunting grounds vanish. As the author notes, her book "explains how our nation nearly squandered the planet's largest temperate rain forest, and how over 20 years a handful of environmentalists, fishermen and scientists succeeded in making the fate of the Tongass an international issue and rescuing this priceless legacy from the chainsaw."

Ecotrust and Conservation International. 1992. *Coastal Temperate Rain Forests: Ecological Characteristics, Status, and Distribution Worldwide.* Occasional Paper Series No. 1. Portland, OR: Ecotrust. 64p.

Ecotrust and Conservation International prepared this document to solicit comments from researchers and managers working in coastal temperate rainforests around the world. The report provides a provisional definition of coastal temperate rainforests and an overview of the subject. It also includes preliminary maps and tables illustrating the "original" extent of coastal temperate rainforests.

Ellis, Gerry, and Karen Kane. Forward by Gaylord Nelson. 1991. *America's Rain Forest.* Minocqua, WI: NorthWord Press. 160p.

Gerry Ellis, world-renowned photographer and naturalist, and Karen Kane, author of many books and audiovisual scripts on nature, are a husband-and-wife team who celebrate the ancient rainforests of the Pacific Northwest in this work. It covers the ancient rainforests that stretch in an "emerald string" from northern California to British Columbia to southeast Alaska's Tongass National Forest, the largest tract of temperate rainforest in the world. Illustrated with 170 color photographs, the book looks at the geology, climate, and biological diversity of the forests, explains how they are interconnected, and describes the indigenous people. The authors call for action to protect America's threatened rainforest.

Herndon, Grace. 1991. *Cut and Run: Saying Goodbye to the Last Great Forests in the West.* Telluride, CO: Western Eye Press. 239p.

Written by an investigative newspaper reporter, this book chronicles forest and logging controversies in the western United States, defining and personalizing the issues. The focus is on U.S. Forest Service timber policies and how those policies have affected the West. The author includes personal interviews with people of each state.

Kelly, David, and Gary Braasch. 1988. *Secrets of the Old Growth Forests.* Layton, UT: Gibbs Smith. 99p.

Without delving into the maze of issues concerning old-growth forests, the authors have produced an informative and stunning look at the ancient forests of the Pacific Northwest. The book's large format and lush photographs help the viewer gain an appreciation for ancient forests. An informative appendix called "One Forest, Many Battlegrounds" describes the various parcels of ancient forest land.

Ketchum, Robert, and Carey D. Ketchum. 1987. *The Tongass: Alaska's Vanishing Rain Forest.* New York: Aperture. 112p.

This large-format book contains beautiful photographs by a landscape photographer of a little known rainforest, the Tongass in Alaska. Brief text provides an introduction to the various threats to the Tongass. The book is not meant to be a study of the rainforest, but rather it is designed to foster awareness of and appreciation for the Tongass.

Kirk, Ruth, with Jerry Franklin. 1992. *The Olympic Rain Forest: An Ecological Web.* Seattle, WA: University of Washington Press. 128p.

Author and photographer Ruth Kirk lived in the Olympic National Forest on Washington's Olympic Peninsula while her husband served as a ranger-naturalist with the Olympic National Park. She coauthored this book with Jerry Franklin, who was chief plant ecologist for the U.S. Forest Service and now is professor of ecosystem analysis at the University of Washington. Together the two have prepared a readable and scientifically sound text that focuses on the Olympic National Forest and its unique ecological web. The book includes line drawings, maps, and 100 color photographs that capture the variety and grandeur of this magnificent forest, which has been recognized as a national park, a World Biosphere Reserve, and a World Heritage Site.

Maser, Chris. 1989. *Forest Primeval: The Natural History of an Ancient Forest.* San Francisco, CA: Sierra Club Books. 282p.

Beginning in the year 988, this book traces the history of an ancient forest in the Cascade Mountain range in Oregon. Chris Maser presents a historical timeline of events that occurred during the lifetime of the mountain trees. Along the way, he also presents the natural history and biology of the flora and fauna of the forest. This volume is a unique comparison of the history of ancient fir trees to the human history taking place at the same time. It includes references, a glossary, and black-and-white photographs.

McAllister, Ian, Karen McAllister, and Cameron Young. 1998. *The Great Bear Rainforest: Canada's Forgotten Coast.* San Francisco, CA: Sierra Club Books. 144 p.

Ian and Karen McAllister spent seven years photographing and mapping the coastal rainforest between Vancouver Island and Alaska in an effort to focus attention on preservation of the forest and ecosystems, threatened by clear-cut logging. The book contains 150 color photographs, including extraordinary images of wild bears. The text is written by the McAllisters, who are founding members of the Raincoast Conservation Society, and environmentalist Cameron Young, an environmental journalist.

Middleton, David. 1992. *Ancient Forests.* San Francisco: Chronicle Books. 107p.

With beautiful photographs and text, David Middleton defines old-growth forests and some of their inhabitants. Although he attempts to present answers to the question about whether old-growth forests should be cut, he does not address the complex issues associated with logging practices in the Pacific Northwest. However, the book accomplishes its purposes: to enhance appreciation for the old-growth forests.

Norse, Elliot. 1990. *Ancient Forests of the Pacific Northwest.* Washington, DC: Island Press. 327p.

Elliot Norse's book includes descriptions of ancient forests and their locations and climate and presents a history of forest use by humans. Norse also discusses the forest ecosystem in detail and introduces the concept of biological diversity. However, the book is essentially an analysis of the U.S. Forest Service's long-term management plans, which Norse concludes are inadequate.

Pynn, Larry. 2000. *Last Stands: A Journey Through North America's Vanishing Ancient Rainforest.* Corvallis: Oregon State University Press.

In this book, award-winning journalist Larry Pynn describes his journey through the coastal temperate rainforests from California to Alaska. Pynn takes an unusual approach in his book, combining his observations of the forest environment (from seabirds to mushrooms to wolverines) with anecdotes about people involved in forest activities. His interviews with loggers, U.S. Forest Service personnel, local government officials, biologists, environmental activists, and many others dramatically show the complex factors at work in the embattled coastal rainforest.

Servid, Carolyn, and Donald Snow, eds. 1999. *The Book of the Tongass.* Minneapolis, MN: Milkweed Editions. 275p.

Thirteen Alaskans contributed to this book, which describes the Tongass National Forest, its wildlife, its economic opportunities, and in two pieces by Tlingit storytellers, its oral history. Readers not only are offered a view deep inside the forest, but also learn about the legislation and economics of the U.S. forestry industry on public lands. The ongoing debate about how the Tongass is to be used is summarized.

Shoaf, Bill. 1999. *The Taking of the Tongass.* Sequim, WA: Running Wolf Press. 288p.

Written by a former U.S. Forest Service employee, this book documents the corruption that for decades sustained the timber sale program in the Alaskan Tongass National Forest. Even though he had never considered himself an environmentalist (or a "greenie," as he puts it), Bill Shoaf quit his job as a forester and became an activist to oppose the Ketchikan Pulp Company, which operated "irrespective of the law and the principles of forestry." As the book describes, his opposition cost him much personal suffering and financial loss and placed his health and safety in jeopardy.

Zuckerman, Seth. 1991. *Saving Our Ancient Forests.* Los Angeles: Living Planet Press. 116p.

Based on research sponsored by the Wilderness Society, this is a story of the ancient forests in the Pacific Northwest. Written in a concise and lively manner, it first defines the ancient forests and then covers such issues as the ecology of the forests, the threats they face from logging, and what can be done about the problem of disappearing forests. The book is illustrated and contains numerous sidebars with "fun facts" about forests.

Tropical Rainforests

Almeda, Frank, and Catherine M. Pringle, eds. 1988. *Tropical Rainforests: Diversity and Conservation.* San Francisco, CA: California Academy of Sciences. 206p.

A review of research and conservation efforts in Costa Rica, this volume is a compilation of papers presented at a 1985 symposium on diversity and conservation of rainforests at the California Academy of Sciences. Two-thirds of the book is devoted to the experiences of the Organization of Tropical Studies and its research in Costa Rica. A good history of the organization and a review of the status of Zona Protectora in Costa Rica also are included.

Arvigo, Rosita, and Michael Balick. 1993. *Rainforest Remedies: 100 Healing Herbs of Belize.* 2nd ed. Twin Lakes, WI: Lotus Press. 336p.

Students of rainforest medicinal plants will find this book an excellent resource, with drawings that make field identification easy. Both the common and scientific names of plants are presented.

Beazley, Mitchell, ed. 1981. *The International Book of the Forest.* New York: Simon & Schuster. 224p.

In its foreword this book is called "an enthralling journey of exploration through the forests and jungles of the six continents . . . the first to examine the beauty, uniqueness, and crucial importance of all the world's forests." Beginning with a discussion of the forest ecosystem—how it works and how people relate to it, the book covers diverse types of forests in Asia, Africa, Australia, Europe, and North and South America, with subsections that deal with the Olympic Rainforest and various tropical rainforests. Excellent photographs, drawings, and maps are included.

Bevis, William W. 1995. *Borneo Log: The Struggle for Sarawak's Forests.* Seattle and London: University of Washington Press. 245p.

This book combines a travel narrative with a serious study of the timber industry that is rapidly destroying the homelands and cultures of the indigenous people of Sarawak, a Malaysian state. Bevis covers the history, economy, and political life of the region along with narratives about Sarawak's people—from traditional peoples to timber company and government officials. A map of Bevis's travel route and black-and-white photographs are included in the book.

Bowman, D.M.J.S. (David). 2000. *Australian Rainforests: Islands of Green in a Land of Fire.* New York: Cambridge University Press. 345p.

The author of this academic study describes it as "a personal journey as much as an academic work on the environmental limits of Australian rainforests." Beginning with a chapter entitled "What Is Australian Rainforest?" this book then discusses rainforest "islands" in Australia, from temperate Tasmania to the northern tropics: their ancient history, how fire started by lightning and other natural causes affected the distribution of rainforests on the continent, and how rainforest conservation depends on fire management. Tables, graphs, and illustrations supplement the text.

Bruenig, Eberhard F. 1996. *Conservation and Management of Tropical Rainforests: An Integrated Approach to Sustainability.* Wallingford, Oxon, UK: CAB International. 339p.

This academic study was written by the chair of world forestry at the University of Hamburg, Germany, who has undertaken research and practical work in forest management since 1954. Because much of Bruenig's work has been in Sarawak rainforests, the book focuses on examples from that area "to demonstrate the universal problems and show how solutions can be approached." An advanced text for students of rainforests, this book is also a manual for rainforest researchers and managers. The book develops a holistic approach to the management and conservation of rainforests, advocating the principles of integrated conservation and management that lead to sustainability. In his preface, the author explains that one of the purposes of his book is to highlight the "large body of knowledge, experience and tradition available to ecology and forestry in the tropical rainforest which is poorly applied," to correct some of the "scientific myths and public misunderstandings of the ecology of the rainforest and of the cultural systems involved," and to "discuss features of the natural forest and socio-cultural ecosystems which can be understood as adaptive mechanisms and can be mimicked in the design of self-sustainable forests that are viable, robust and tolerant, capable of coping flexibly with uncertainty." Numerous figures and tables and a helpful list of acronyms, abbreviations, and symbols are included in the book. A glossary and forest technology terms are appended as well.

Caufield, Catherine. 1986. *In the Rainforest: Report from a Strange, Beautiful, Imperiled World.* Chicago: University of Chicago Press. 306p.

This book has been read and studied extensively by those involved in rainforest preservation. It describes rainforest regions worldwide and explains the political and economic causes for rainforest destruction.

Dalton, Stephen, George Bernard, and Andrew Mitchell. 1990. *Vanishing Paradise: The Tropical Rainforest.* Woodstock, NY: The Overlook Press. 176p.

The work of Stephen Dalton and George Bernard, nature photographers who are known worldwide, make up this large-format book. Many of the 200 dramatic and stunning photographs show tropical forest animals, birds, and plants in exquisite detail. Some photographs were taken in Costa Rica's rainforest. Zoologist Andrew Mitchell wrote the text, which describes the methods by which the tropical rainforest is able to regulate moisture in the air, the variety of life in the forest, and how animals find food, hunt prey, and reproduce. A chapter on communication shows how mammals and birds stay in touch in a rainforest by being "very noisy, very colourful, or very smelly." An especially interesting section shows how mammals, birds, and insects are able to camouflage themselves. Other sections explain the forest canopy and how decay and renewal work hand-in-hand so that the forest can recycle itself.

Dixon, Anthony, Hannah Roditi, and Lee Silverman. 1991. *From Forest to Market: A Feasibility Study of the Development of Selected Non-Timber Forest Products from Borneo for the U.S. Market.* Cambridge, MA: Cultural Survival. 2 vols., 300p total.

Working with Cultural Survival, three Harvard Business School students carried out the studies that led to this report on how a business framework can be used to develop and market nontimber forest products from the rainforests of Borneo. Written as handbooks, the publications contain descriptions of products and their uses and strategies for their development, information on trade regulations, ideas about how to influence policy makers, and lists of individuals and groups working for business alternatives to forest destruction. Color photographs of ninety products are included.

Downing, Theodore E., Susanna B. Hecht, Henry A. Pearson, and Carmen Garcia-Downing, eds. 1992. *Development or Destruction: The Conversion of Tropical Forest to Pasture in Latin America.* Boulder, CO: Westview Press.

Introducing this work, the editors point out that "[t]he conversion of forest to grasslands in the humid tropics is one of the most profound land transformations of the 20th Century, with major consequences for biodiversity, global and regional climates, soil resources, and local populations." Contributors in the first section explain the livestock economy and how it relates to forest destruction. In the following sections, contributors explain the environmental and social impacts of raising livestock in rainforest regions and alternatives to livestock production. Also included are the views of community representatives, peasants, and rubber tappers.

Dwyer, Augusta. 1990. *Into the Amazon: The Struggle for the Rain Forest.* San Francisco, CA: Sierra Club Books. 250p.

In this wide-ranging book on Amazonia, Augusta Dwyer discusses the politics of the region, the struggles of Chico Mendes, and the myths and legends of the land. The author presents a passionate account of a vanishing way of life in Amazonia, describing how government, industrialists, and international exploiters have destroyed or are seeking to destroy an area of the world that is essential to the planet's survival.

Forsyth, Adrian, Michael Fogden, and Patricia Fogden. 1990. *Portraits of the Rainforest.* Ontario, Canada: Camden House. 156p.

A biologist and natural-history writer, Adrian Forsyth presents a collection of essays on diverse aspects of the tropical rainforest in this large-format book. Chapters describe the origins of varied plants and animals, explain why rare species are common in the tropics, define the nutrient cycles, and describe the roles of diverse life-forms—from rainforest amphibians to wood-eating termites to the "ultimate predator," the jaguar. Color photographs by the Fogdens enhance the text throughout.

Forsyth, Adrian, and Ken Miyata. Foreword by Dr. Thomas Lovejoy. 1995. *Tropical Nature: Life and Death in the Rain Forests of Central and South America.* New York: Simon & Schuster. 248p.

This is a readable biological treatise about basic tropical ecology. A beginning chapter explains how climate dictates all else in the tropics. Other chapters cover the natural history of various plants and animals.

Gallant, Roy A. 1991. *Earth's Vanishing Forest.* New York: Macmillan. 162p.

Written for young adults, this introductory book on the importance of the rainforest describes the forest's biodiversity and the threat to the homelands of indigenous people. Roy Gallant also explains how the destruction of forests leads to the loss of medicinal plants and contributes to the buildup of "greenhouse gases" in the atmosphere. He concludes that time is running out, that the secrets of the rainforest could be lost forever. The book is illustrated with black-and-white photos and includes maps and a glossary.

Gennino, Angela, ed. 1990. *Amazonia: Voices from the Rainforest.* San Francisco, CA: Rainforest Action Network. 92p.

A highly recommended reference, this paperback guide provides profiles of 250 international organizations working on Amazonia issues. These include groups active in Latin America, Europe, North America, and the Asia/Pacific area. The reference also includes a bibliography of recommended books and films on Amazonia. It is a good resource for anyone interested in the politics of deforestation and efforts worldwide to stop the destruction of rainforests.

Gentry, Alwyn H. 1990. *Four Neotropical Rainforests.* New Haven, CT: Yale University Press. 627p.

This collection of papers was the result of a 1987 symposium on neotropical rainforests. All of the papers are based on a comparison of ecosystem dynamics at four research stations: LaSelva in Costa Rica, Barro Colorado Island in Panama, Cocha Cashu in Peru, and Manaus in Brazil. It includes an extended section on forest dynamics.

Goulding, Michael. 1980. *The Fishes and the Forest: Explorations in Amazonian Natural History.* Berkeley and Los Angeles: University of California Press. 280p.

The premier fish naturalist of the Amazon system, Michael Goulding has worked in the Amazon since 1973 and has explored more than fifty rivers. This is a seminal work on Amazonian fish and their dependence on the vast seasonally flooded forest.

Goulding, Michael, Nigel J.H. Smith, and Dennis Mahaar. 1996. *Floods of Fortune: Ecology and Economy along the Amazon.* New York: Columbia University Press. 193p.

The authors of this important book are experts on Amazonia. Michael Goulding is director of the Amazon Rivers Program of the Amazon Conservation Association. Nigel Smith is professor and chair of the Department of Geography at the University of Florida and previously was a senior researcher at Worldwatch Institute and a research associate at Brazil's National Institute for Amazon Research. Dennis Mahar is a visiting lecturer in the Department of Economics at the University of Florida, and he has been chief of the World Bank's Environmental Division for Latin America and the Caribbean. The book describes the stunning diversity of plant and animal life found in and along the rivers, and it shows how human populations from earliest times to the present have used the rivers and floodplains. It explains the complex environmental and social problems associated with Amazonian development and offers solutions that reconcile development with conservation. Nearly one hundred color photographs enrich the work.

Gradwohl, Judith, and Russell Greenberg. 1988. *Saving the Tropical Forests.* Washington, DC: Island Press. 214p.

Released in conjunction with a major Smithsonian Institution traveling exhibition, this book examines the deforestation of tropical rainforests and offers immediate and timely solutions in the form of case studies. The case studies, based on projects that have been implemented throughout the developing world, are short and informative—meant to be readable, not to provide a thorough analysis.

Griffiths, Michael. 1990. *Indonesian Eden: Aceh's Rainforest.* Baton Rouge: Louisiana State University Press. 111p.

Michael Griffiths, who spent many years traveling through the Indonesian rainforests, presents his travels in a journal format with lush color photographs. This large-format book includes photos of rare animals and interviews with people who live in the rainforest.

Head, Suzanne, and Robert Heinzman, eds. 1990. *Lessons of the Rainforest.* San Francisco, CA: Sierra Club Books. 256p.

Essays from twenty-four leading authorities on the effects of worldwide destruction of tropical rainforests are included in this challenging book. In a global context, the essays show how international economic and social patterns have affected the well-being of rainforests. They cover such topics as "Tropical Forests and Life on Earth," "Five Hundred Years of Tropical Forest Exploitation," "Asia's Forest Cultures," "Extinction," and "Saving the Forest and Ourselves." The experts warn that a variety of reforms are needed to "ensure the sustainability of life on our planet" and suggest ways to reach that goal.

Hecht, Susanna, and Alexander Cockburn. 1989. *The Fate of the Forest: Developers, Destroyers and Defenders of the Amazon.* London and New York: Verso. 266p.

This book traces the history of the Amazon rainforest over four centuries, explaining how explorers, conquerors, naturalists, and entrepreneurs have viewed the riches of the Amazon. The study describes the realm of nature, the "heritage of fire" in the forest, the exploitation of resources, and the many ways the region is being devastated through the destruction of flora and fauna, the poisoning of rivers, and the persecution and killing of rubber tappers and settlers. A major section of the work covers the events leading up to the 1988 murder of Chico Mendes, leader of the rubber tappers. A Forest People's Manifesto, with policies for agrarian reform and a call for an end to the oppression of forest people, is appended. As the authors of the book point out, the rainforest will not flourish unless the people who live there are able to prosper, too. There is an extensive bibliography and a map showing Amazon reserves as of early 1989.

Holm-Nielsen, L. B., L. C. Nielsen, and H. Baslev, eds. 1989. *Tropical Forests.* San Diego, CA: Academic Press. 374p.

This book is a collection of technical and scientific papers that were presented at a symposium held at Aarhus University in Denmark. At the symposium, titled "Tropical Forests: Dynamics, Speciation and Diversity," experts presented studies conducted in Sri Lanka, Guiana, Borneo, China, and Amazonia. The editors of this work note in the preface that a baseline study needs to be completed on taxonomies and species before any new under-standing of the tropical ecosystem can be articulated. Charts, graphs, and an index of plant names are included.

Hurst, Philip. 1990. *Rainforest Politics: Ecological Destruction in Southeast Asia.* Atlantic Highlands, NJ: Humanities Press International. 320p.

This well-documented book describes the destruction of rain-forests in Southeast Asian countries: Thailand, Indonesia, Malaysia, and the Philippines. Philip Hurst explains the political and social aspects of deforestation and the conflict between the industrialized and nonindustrialized countries.

Jacobs, M., and R. A. A. Oldeman. 1988. *The Tropical Rain Forest: A First Encounter.* Translated by R. Kruk. New York: Springer-Ver-lag. 311p.

This book is a thorough overview of the biological and human as-pects of tropical rainforests. It covers the development of public awareness about rainforest issues and explains how rainforests are studied. Chapters discuss such topics as rainforest climate, soil, plants, and animals. Several chapters focus on the rainforests of specific areas (tropical Africa, for one). The value of rainforests, the damage and destruction they suffer, and efforts to protect these ecosystems are also covered. The book includes 170 illustrations.

Johnson, Brian. 1991. *Responding to Tropical Deforestation.* Balti-more, MD: WWF Publications. 63p.

This book is part of a series of research papers published by the World Wildlife Fund. The report contains an overview of the causes and consequences of the tropical rainforest crisis and an-alyzes three treaty negotiations and other agreements promoted by the United Nations conferences on Environment and Devel-opment. The author proposes fully using existing institutions rather than developing new ones.

Jordan, Carl F. 1985. *Nutrient Cycling in Tropical Forest Ecosystems: Principles and Their Application in Management and Conservation.* New York: John Wiley and Sons. 180p.

As the title suggests, this book is a technical work aimed at foresters and forestry students. Chapters include "Factors Which

Control Nutrient Cycles," "Nutrient Conserving Mechanisms," "Differences in Ecosystem Characteristics along Environmental Gradients," "Characterization of Nutrient Cycles," "Changes in Nutrient Cycles Due to Disturbance," and a summary of recommendations for managing tropical ecosystems. Charts, graphs, and an extensive bibliography are included.

Kane, Joe. 1989. *Running the Amazon*. New York: Alfred A. Knopf. 227p.

In this true-life account, Joe Kane describes the first expedition to ever travel the full length of the Amazon River. Kane, who was the only American in the multinational crew, tells the story of how the crew learned to live together and to accept the challenges of the white waters of the Amazon during the six-month journey. He also provides detailed descriptions of the land and river as well as the adventures of the expedition.

Kimerling, Judith, with the Natural Resources Defense Council. 1991. *Amazon Crude*. Washington, DC: The Natural Resources Defense Council. 132p.

The devastating effects of virtually uncontrolled oil exploration in the Ecuadorean rainforests are the subject of this well-documented book. Not only are forests destroyed, but rainforest people are poisoned, debilitated, and killed by toxic pollutants from oil drilling and diseases introduced by outsiders. Full-color photographs, maps, and illustrations are included.

Kricher, John. 1990. *A Neotropical Companion: An Introduction to the Animals, Plants, and Ecosystems of the New World Tropics*. Princeton, NJ: Princeton University Press. 436p.

Written for the layperson, this overview of neotropical forests would be of interest to any armchair traveler. John Kricher defines rainforests, explains their functions and ecosystems, and briefly describes the medicinal plants and the many animals, birds, and reptiles found there. A glossary is included.

Lamb, F. Bruce. 1985. *Rio Tigre and Beyond: The Amazon Jungle Medicine of Manuel Cordova*. Berkeley, CA: North Atlantic Books. 227p.

Picking up from his earlier book, *Wizard of the Upper Amazon*, Lamb in this book completes the life story of Manuel Cordova, whom he calls "a true bi-cultural healer." A Peruvian rubber worker, Cordova was kidnapped by indigenous people and trained for seven years to be their new shaman, learning how to use plants as medicinals and to develop psychic powers, which he used extensively for diagnostic purposes. Anecdotal in style, this book is also a story of the healing properties of medicinal plants still in use today.

Lewis, Scott. *Rainforest Book.* 1990. Venice, CA: Living Planet Press. 112p.

Chock-full of facts about rainforest animal and plant species, this book is written in a layperson's terms. It clearly explains why species are threatened and what can be done to save them. The last two chapters list resources and organizations working to prevent deforestation in the tropics.

Margolis, Mac. 1992. *The Last New World: The Conquest of the Amazon Frontier.* New York: W.W. Norton & Company. 367p.

Although this is a story about the waste and ruin that has been the result of South America's version of Manifest Destiny, it is also the story of peasants, cattle ranchers, and rubber tappers who are trying to find ways to develop land areas without destroying the region's complex ecosystems. Mac Margolis shows that with the help of scientists and extension workers, people in the Amazon region are trying to find a middle course—between total preservation and total destruction. The book is illustrated with black-and-white photographs.

Martin, Claude. 1991. *Tropical Rainforests of West Africa.* Basal, Switzerland: Birkhauser Verlag; Boston: Birkhauser Boston. 235p.

During the 1980s, approximately 7,200 square kilometers of rainforest were destroyed each year in West Africa. Arguing that the rainforests in West Africa are the forebears of what has happened and will happen to the rainforests throughout the world, Martin discusses the history, ecology, conservation, and future of West Africa's rainforests. One of Martin's main contentions is that scientists have failed to set research priorities and to explore the complexity of the tropical forest ecosystem. This book, one of the few available on West African rainforests, includes descriptions of many species of primates and small mammals that live in the forests. It also explains conservation measures that are being taken. Color photographs, charts, graphs, maps, and an extensive bibliography are included.

McIntyre, Loren. 1991. *Amazonia*. San Francisco, CA: Sierra Club Books. 164p.

This large-format book is a collection of 121 lush color photographs by Loren McIntyre, a photojournalist who has spent years photographing South America's wilderness. The structure of the book and section titles correspond to different parts of the Amazon river. Photos show the Amazon Basin and the "white water" of the mountainous region in the west, the "black water" of the Rio Negro in the north, and the "brown water" and "blue water" sections to the south and east respectively. Each section includes photos of plant and animal life and human culture, with the author's insights in text form interwoven throughout.

Miller, Kenton, and Laura Tangley. Introduction by Gus Speth. 1991. *Trees of Life: Saving Tropical Forests and Their Biological Wealth*. Boston: Beacon Press. 228p.

Written for scientists and laypersons alike, this book is part of the World Resources Institute Guide to the Environment series. The authors compare the historical treatment of North American forests to what is occurring now in Amazonia, and they focus on efforts to prevent deforestation. A chapter on old-growth forests is also included along with seven pages of forest facts.

Mitchell, Andrew. 1986. *The Enchanted Canopy: A Journey of Discovery to the Last Unexplored Frontier, the Rainforests of the World*. New York: Macmillan. 288p.

Written for a general audience as well as the informed layperson, this book includes ninety-eight color photographs of tropical rainforest scenes, with interwoven text based on research conducted in rainforest regions worldwide. A major portion of the book explains and shows tree-top ecosystems, including photos of insect life and predator-prey relationships unseen from the forest floor. The book also provides information on the people who live in close association with the rainforest canopy.

Mitchell, George J. 1991. *World on Fire: Saving an Endangered Earth*. New York: Macmillan. 320p.

While a U.S. senator, George Mitchell was one of the leading environmentalists in the U.S. Congress and was chair of the Committee on Environmental Protection. Written while he was still a member of Congress, his book examines global ecological crises, including changing climate due to the greenhouse effect, overpopulation, and deforestation. He also describes the steps that must be taken to prevent major catastrophes. Mitchell explores the U.S. role in developing global conservation strategies, such as saving energy and shifting funds from military spending to environmental preservation.

Myers, Norman. 1992. *The Primary Source: Tropical Forests and Our Future.* New York: W.W. Norton & Company. 416p.

A well-known author and environmental consultant, Norman Myers clearly explains that tropical forests have been destroyed because of the profit incentives of multinational companies, the severe economic needs of developing nations and overconsumption in developed nations. In this work, which was first published in 1984 and has since been updated, Myers offers a blueprint for solving some of these problems. In this edition, Meyers provides information about the condition of the tropical forests and suggests new geopolitical strategies for their survival. The book includes references, tables, and black-and-white photographs.

Newman, Arnold. 1990. *Tropical Rainforest.* New York: Facts on File. 256p.

A conservationist and science writer who has spent many years traveling to rainforests and studying their ecology, Newman brings together much detailed information in this illustrated work. It is divided into five sections: "What Is a Tropical Rainforest?" "The Web of Life," "Threats to the Forest," "What Do We Lose?" and "A Blueprint for Survival." With text and photos, this is a comprehensive look at tropical rainforests and includes helpful appendices listing educational resources, tropical timbers and domestic alternatives, and an essay entitled "Toward Sustained Productivity." Several hundred entries are listed in the extensive bibliography.

O'Connor, Geoffrey. 1997. *Amazon Journal: Dispatches from a Vanishing Frontier.* New York: Dutton. 256p.

Documentary filmmaker Geoffrey O'Connor reports on the little-known quest for gold in Amazonia, which adversely affects indigenous people. He describes not only disputes over gold but also the differences between the culture of indigenous people and that of the industrialized world. Those differences, O'Connor writes, resulted in tragedies that could have been avoided if there had been better understanding of cultural barriers.

Payne, Junaidi, with photographs by Gerald Cubitt. 1990. *Wild Malaysia: The Wildlife and Scenery of Peninsular Malaysia, Sarawak, and Sabah.* Cambridge, MA: MIT Press. 208p.

The rainforests of Malaysia in Southeast Asia were undisturbed just a century ago, but today the thirteen-state federation supplies more tropical wood products than any other region in the world. With more than 400 color photographs by a leading natural history photographer, this book shows the flora and fauna of Malaysia. Payne, who is project director of World Wide Fund Malaysia, describes the history and culture of Malaysia's people and how they are trying to prevent degradation of their environment.

Perlin, John. 1989. *A Forestry Journey: The Role of Wood in the Development of Civilization.* New York: W.W. Norton & Company. 445p.

In this book, John Perlin traces the role of wood in society, from the Bronze age to the nineteenth century, showing the role forests have played in the development of past civilizations. He explains that because wood was the primary fuel and building material in the past, its abundance or scarcity shaped the culture, demographics, economy, politics, and technology of each society. By examining how other societies dealt with the consequences of deforestation, Perlin provides insights that help people today understand and resolve problems associated with forest destruction.

Perry, Donald. 1986. *Life Above the Jungle Floor.* New York: Simon & Schuster. 170p.

Two-thirds of animal and plant life in the rainforest resides in the canopy. In this volume, Donald Perry, a pioneer in the exploration of rainforest canopies, describes the rainforest canopy of Costa Rica. He also explains how he devised a working area with a web of ropes and platforms from which he could study the biome. Color photographs illustrate the book.

Prance, Ghillean T. 1998. *Rainforests of the World: Water, Fire, Earth, and Air.* London: Havill Press. 298p.

Sir Ghillean Prance, a highly acclaimed naturalist, wrote the text for this work, which is illustrated with more than 200 color photographs by world-renowned photographer Art Wolfe. Together the text and photographs present an intimate view of the rainforest and its inhabitants—the plants, animals, and indigenous peoples that make their homes in the most beautiful and endangered places on earth. The book is divided into four sections—"Water," "Fire," "Earth," and "Air"—and explores the different ecological workings of the rainforests.

Revkin, Andrew. 1990. *The Burning Season. The Murder of Chico Mendes and the Fight for the Amazon Rain Forest.* Boston: Houghton Mifflin. 298p.

This book is the story of Chico Mendes, a rubber tapper and environmental activist in Brazil who was murdered by wealthy cattle ranchers intent on clearing the forest and using the land for cattle grazing. The murder of Mendes brought cries of outrage and protest from around the world, and Mendes's name now symbolizes global efforts to save rainforests. Andrew Revkin describes the events that led up to the murder, focusing on how Mendes and his fellow rubber tappers organized to preserve the forest upon which they depend for survival. He also describes the natural alliance between the tappers and the indigenous tribal groups, all of whom need the rainforest to live.

Richards, Paul W., with contributions by R.P.D. Walsh, I.C. Baillie, and P. Greig-Smith. 1996. *The Tropical Rainforest: An Ecological Study.* 2nd ed. New York: Cambridge University Press. 575p.

In this revised edition, Professor Richards provides a personal view of rainforest ecology, based on sixty years in the field. This book covers such topics as the structure of the primary forest, regeneration, trees and shrubs, reproductive biology, soils of the humid tropics, and the composition of primary rain forests. A postscript describes the future of the tropical rain forest, and an index of plant names is included.

Rifkin, Jeremy. 1992. *Beyond Beef: The Rise and Fall of the Cattle Culture.* New York: Dutton. 353p.

When Rifkin wrote this book, he noted that there were "1.28 billion cattle populating the earth. They take up nearly 24 percent of the landmass of the planet and consume enough grain to feed hundreds of millions of people. . . . Cattle raising is a primary factor in the destruction of the world's remaining tropical rain forests." One underlying theme of this book is that the growing world cattle industry is a major contributor to the ecological devastation in many areas of the world, particularly in forestlands that have been cleared for cattle production. Cattle are produced "at the expense of a burned forest, an eroded rangeland, a barren field, a dried-up river or stream, and the release of millions of tons of carbon dioxide, nitrous oxide, and methane into the skies," notes Rifkin, an environmental activist. His book includes statistics on cattle production—the fourth largest manufacturing industry in the United States, using over 70 percent of the grain produced for cattle feed. The book is also a call for reassessment of the effects of grain-fed cattle production worldwide.

Silcock, Lisa, ed. 1990. *The Rainforests: A Celebration.* San Francisco, CA: Chronicle Books. 223p.

This elegant large-format book, with a foreword by Prince Charles, is a pictorial celebration of the world's rainforests compiled by an international team of experts and photographers. Each chapter begins with an overview of a particular web of life in the forest, and then a strand from that web—a plant, an insect, a bird, a reptile, or a mammal—is illustrated and described. The last chapter describes the people who make the forest their home and are being threatened by deforestation. Many of the color photographs show aspects of the forest that have never been photographed before, enhancing the the reader's appreciation of the rainforest.

Shoumatoff, Alex. 1990. *The World Is Burning: Murder in the Rain Forest.* Boston: Little, Brown. 377p.

In his analysis of deforestation in the Amazon, excerpts of which have been published in *Vanity Fair,* Alex Shoumatoff tells the story of the 1988 murder of Chico Mendes, a grassroots activist and leader of a rubber workers' organization in Amazonia. Shoumatoff places this story against a background of the history, economics, and ecology of the region, presenting a firsthand account of some of the destruction and cruelty perpetrated by those who exploit the rainforest.

Terborgh, John. 1992. *Diversity and the Tropical Rain Forest.* NY: W. H. Freeman and Company. 242p.

For the general reader, this colorfully illustrated Scientific American Library book includes interesting, readable text on ecological theory and conservation. The book emphasizes the importance of each part of the forest, from the towering canopy of emerging trees to the microscopic fungi on the forest floor.

———. 1999. *Requiem for Nature.* Washington, DC: Island Press. 234p.

In this book John Terborgh, a professor of environmental science and botany at Duke University, estimates that at the present rate of rainforest destruction, the last tree in the last primary forest will fall in 2045. Terborgh has spent half his time over more than two decades researching in Manú National Park in Peru, experience that has yielded his unique perspective on the role of parks, societies, and concerned institutions in preserving nature. The author does not hold much hope for saving the world's rainforests and biodiversity without a vast change in public opinion. As he bluntly puts it: "Whether we like it or not, tropical forests are worth more dead than alive." Still, he offers some of his advice for preservation, including national conservation trust funds and strict policing of forest preserves.

Whitmore, Timothy C. 1998. *An Introduction to Tropical Rain Forests.* 2nd ed. New York: Oxford University Press. 226p.

Written by a senior researcher at the Oxford Forestry Institute, this book is primarily an introductory text on tropical rainforests for the general college student. It also provides information that can be easily understood by a broad readership. Chapters describe the climate, formations, and growth cycles in the forests; plant life and seasonal rhythms; the diversity of rainforest animals; and the interconnections between plants and animals. The text also covers tropical rainforests through time—plant and animal distribution patterns—from geological periods millions of years ago to the present. Diagrams, charts, maps, and black-and-white photographs supplement the text. Although many scientific terms are defined within the text, a glossary is a helpful addition, as is the "Index to Plants and Forest Products."

Selected Nonprint Resources

Nonprint resources selected for this section include audio books, bibliographies and databases on the World Wide Web, computer programs and CD-ROMs, films and videos, filmstrips, and videocassettes. Most of these resources were produced between 1990 and 2000, and companies that offer them for rent or purchase are listed in the entries as "Sources." Addresses for these companies and for publishers of the print resources are found at the end of the chapter.

Audio Books

**Environmental Overkill: Whatever Happened
to Common Sense**
Author: Ray, Dixie Lee
Type: Cassettes (6 tapes)
Source: Blackstone Audiobooks

This book, read by Jeff Riggenbach, was first published in print
form in 1993 by Regnery. It was written by the former governor
of Washington, Dixie Lee Ray, who has often criticized environmentalists, claiming they are threatening property rights. In
Ray's view, "environmental extremists, scare-mongering journalists," and "media-conscious scientists" are among those attempting to destroy competition and economic development.

Savages
Author: Joe Kane
Type: Unabridged cassettes (8 tapes)
Source: Books on Tape

Read by Michael Russotto, this book by Joe Kane was published
in print form in 1995 by Alfred A. Knopf. An American journalist, Kane tells the story of the Huaorani, a nation of about 1,300,
with whom he lived for months in the deepest part of the Amazonian rain forest in Ecuador. Although the Huaorani have been
exposed only recently to the modern world, oil production in the
Amazon has opened the forest to colonization and industrialization. Kane reports on the tribe's connections with the old world
and its battles with the new one.

Bibliographies and Databases on the World Wide Web

"Environmental Resources"
http://egj.lib.uidaho.edu/egj02/link01.html
"Green Bibliography"
http://green.ca/english/resource/reading.htm
"An Index of Rainforest Related Sites on the Web"
http://rain-tree.com/links.htm
"Rain Forest Bibliography"
http://dcepea.harvard.edu/extension/cscie23/public_html/im
ages/sharris2/rfinfo.htm
"Rainforest Education.com"
http://www.rainforesteducation.com/

"Rainforest Live Bibliography"
http://www.sitesalive.com/admin/rlbiblio.htm
"Resources for Registered Classrooms Index"
http://www.pbs.org/tal/costa_rica/res2/
"A Temperate Rainforest Bibliography"
http://kids.osd.wednet.edu/Marshall/homepage/biblio.html
"Trees for the Future Bibliography"
http://www.treesftf.org/bibliog.htm

Computer Programs and CD-ROMS

A Guide to the Global Environment
Type: CD-ROM
Source: World Resources Institute

A database on global conditions and trends, this time-saving re-search and reference tool was produced by the World Resources Institute. It contains all the vital economic, population, natural resource, and environmental statistics found in the print edition of *World Resources 1998–1999,* plus a twenty-year time series for many variables. With the software, a person can browse, select, and export any or all the data.

Rainforest Explorer
Type: CD-ROM
Source: Multi Dimensional Communications, Inc.

Hundreds of plant and animal species are brought to life with full-motion, full-screen video. Learn about rare and endangered species like the beautiful ocelot and the gentle tapir. Discover how the indigenous people of the rainforest live and how their lifestyles have been affected by civilization. The intuitive inter-face makes it easy to get from one section to another. Pick from dozens of movies and slide shows. This fully narrated CD-ROM also has a search feature that lets you quickly find subjects, ani-mals, or plants that interest you. Full-motion, full-screen video requires no additional hardware.

Total Amazon
Type: CD-ROM
Source: Phoenix/BFA/Coronet/MTI

Produced by Simon & Schuster Interactive and Computer Cur-riculum, *Total Amazon* is a guide to the diverse lands, waterways, and ecosystems of Amazonia and the South American tropical rainforest. The CD-ROM is geared for grades 7–9, and includes

30 minutes of video, 150 audio passages, 832 articles, and 660 full-color graphics.

Films and Videos

Amazonia: The Road to the Edge of the Forest
Type: Video (in two parts)
Run time: 48 minutes each
Source: Filmakers Library

Produced by the Canadian Broadcasting Company, this production from the acclaimed *Nature of Things* series explains the failure of a massive resettlement program in Brazil, which lured millions of migrants to the Amazon. The settlers cleared the forest with the hope of farming the land, only to discover that the soil was thin and infertile. Despite this, the Brazilian government continued to promote settlement, which threatens many indigenous Amazonian people and rubber tappers. The film includes one of the last interviews with the late Chico Mendes, leader of the rubber tappers; Mendes was murdered for his outspoken opposition to land clearance.

Amazonia: Voices from the Rain Forest
Type: Video
Run time: 70 minutes
Source: The Video Project

This award-winning video gives voice to the native people of the Amazon rainforest as well as riverine dwellers, rubber tappers, and small-scale farmers. All of these groups depend for survival upon the forest along the Amazon River. First-person accounts of the struggle for survival blend with photography of the forest and original music. The film is recommended for high school and adult audiences. A ninety-two-page resource book is included.

America's Rainforest: A Disappearing Glory
Type: Video
Run time: 35 minutes
Source: Films for the Humanities

A Cambridge Educational Production, this program is designed to help students gain an appreciation for the delicate nature of the ecosystems of the old-growth forests on Washington's Olympic Peninsula. Because of commercial timbering practices over the past 300 years, the soil the forest relies on has been degraded and ecosystems have been disrupted. According to the program description, "the beauty, complexity, and wonder of these rapidly disappearing biospheres is thoroughly explored, and the detrimental global effect on the environment is discussed."

Ancient Forests
Type: Video
Run time: 25 minutes
Source: National Geographic

This award-winning video for grades 7–12 takes the viewer along the Pacific coast, from the Tongass National Forest in Alaska to northern California where ancient trees are found. It tells about the conflict over the fate of these vanishing forests.

Arrows against the Wind
Type: Video
Run time: 52 minutes
Source: Bullfrog Films

Filmed secretly in Irian Jaya in West Papua New Guinea, this documentary depicts what is known as the "Amazon of Asia," a land of vast jungles and tropical rainforests. It is an area where, for 25,000 years, numerous indigenous tribes have lived in spiritual harmony with the land. It is the story of two tribes, the Dani and Asmat, and their social, political, and environmental upheaval, which began in 1963 when Indonesia invaded and opened up the land for international development. It has been highly recommended for high school and college audiences and for other adults interested in the survival of indigenous peoples.

Banking on Disaster
Type: Video
Run time: 78 minutes (in three parts for classrooms)
Source: Bullfrog Films

Filmed over a ten-year period, this documentary shows the destruction of the Amazon rainforests. It documents the disastrous consequences of cutting and paving a road through the world's largest rainforest. The program presents several points of view, including those of indigenous people and industry and government officials.

Battle for the Trees
Type: Video

Run time: 57 minutes
Source: The Video Project

Produced by Sarus/Otmoor Productions with the National Film Board of Canada and Channel Four Television of the United Kingdom, this film "documents one of North America's most important and contentious environmental crises—the clearcutting of the largest remaining temperate rainforest on the continent." As the video description explains, "A handful of large, mainly multinational timber companies are clearcutting this ancient forest, located in coastal British Columbia, at an unprecedented rate, causing worldwide concern. Most of the wood being cut is consumed in the United States, where many leading newspapers and phone books are printed on pulp from B.C.'s old growth forests." Massive clear-cuts are shown, along with "often passionate commentary from both sides, including forestry experts, logging company officials, environmentalists, and native people whose way of life is threatened by the clearcuts."

Biodiversity: The Variety of Life
Type: Video
Run time: 42 minutes
Source: Bullfrog Films

For those unfamiliar with the term "biodiversity," this video explains the interconnectedness of life and illustrates how any action taken in an ecosystem affects all of its components. Even though it focuses on one ecosystem, the film presents concepts that apply everywhere. A study guide is available.

Blowpipes and Bulldozers
Type: 16mm color film
Run time: 60 minutes
Source: Bullfrog Films

This highly acclaimed film tells the story of a unique tribe of rainforest people called the Penan, who have lived for over 40,000 years in Sarawak, Borneo, part of Malaysia. Bruno Manser from Switzerland spent five years with the Penan, adapting to their lifestyle and helping to bring in an Australian film crew to document the struggle to save their land and way of life from logging companies supported by the Malaysian government.

Burning Rivers
Type: Video

Run time: 28 minutes
Source: The Video Project

Set in Guatemala, this video shows how environmental and social crises develop when rainforests are chopped down, badly polluted rivers burn, and dangerous pesticides poison farm-workers. These serious environmental problems stem largely from the unequal distribution of wealth and land in Guatemala. The film provides evidence that wealthy property owners use the best land for growing crops to export, leaving little productive land for struggling peasants to feed their families. Some Guatemalans have organized for economic and environmental justice despite severe repression and death threats. A thirty-two-page discussion guide is included.

Can Tropical Rainforests Be Saved?
Type: Video
Run time: 118 minutes
Source: Richter Productions

In this film, rainforest conservation is examined in Cameroon, Rwanda, Brazil, Indonesia, the Philippines, Malaysia, Ivory Coast, Costa Rica, Panama, and Madagascar. Included are discussions on the reasons for deforestation and the advantages of preserving the rainforests. Alternative and sustainable uses of the rainforest are also considered, as well as the effects of deforestation on the indigenous people.

Cry of the Forgotten Land
Type: Video
Run time: 26 minutes
Source: The Video Project

The forgotten land depicted in this film is in Indonesian New Guinea (Irian Jaya), where one of the world's greatest remaining rainforests is located. This rainforest has been home to the indigenous Moi people for more than 2,500 years, but their land is rapidly being destroyed by massive logging, which also threatens 100,000 species found nowhere else in the world. "Filmed at great personal risk, this video is a well-crafted study of both these unique rainforests and the ancient Moi culture, as well as the seemingly overwhelming threats they face. It is a valuable and timely examination of one of the most important contemporary struggles for rainforest preservation," according to the video description.

The Decade of Destruction
Type: Video (single tape or six separate tapes)
Run time: 85 minutes total
Source: Bullfrog Films

This series of six programs, created with support from the World Wide Fund for Nature, is based on the destruction of the Amazonian rainforest during the 1980s. For use in classrooms, the series gives students a well-rounded view of the many complex factors that led to what may be this century's worst environmental disaster. Titles in the series cover topics including the mechanics of the rainforest, the colonists who invaded the rainforest, development roads that led to forest destruction and displaced Amazon Indian tribes and rubber trappers, and politicians who plunder the rainforest.

Equatorial River: The Amazon
Type: 16mm film
Run time: 23 minutes
Source: Bullfrog Films

This film, produced by the National Film Board of Canada, shows how water and nutrient cycles work in the Amazon basin. The Amazon River, the Earth's largest river system, flows through the Amazon rainforest, where most of the known plant species are found, plus numerous animals and insects. The interrelationship of elements in this vast ecosystem is the primary focus of this film.

Forest for the Future
Type: Video (three-volume series)
Run time: 23, 27, and 21 minutes
Source: The Video Project

This series of three videos presents the story of the Pacific Northwest old-growth forests, showing the interconnections within the forest community. The first video, *The Natural Forest*, depicts a functioning ancient forest ecosystem, with scientists explaining the interdependent web of forest species. *Humans in the Forest* is the second video, which looks at the historic and ongoing impact of human activities on old-growth forests. The third video, *Decisions for the Future*, examines ways to balance human activities and the natural system. A 20-page study guide is included for use in classrooms, grade 7 on up. Adults will find the series educational also.

The Forest through the Trees
Type: Video
Run time: 58 minutes
Source: The Video Project

Produced by former NBC News bureau chief Frank Green, this film focuses on the last remaining stands of virgin redwoods in the United States and the endangered spotted owl. During the film, loggers, timber company officials, politicians, environmentalists, and local residents present their views and concerns. Besides the competing interests of these various people, the film explains alternatives to current logging practices and policies regarding natural resources in the United States.

Forestry and Food Security
Type: Video
Run time: 15 minutes
Source: North American and Caribbean Regional Center,
             United Nations Food and Agriculture Organization

This video explains how directly and indirectly, forests play an important nutritional role that is often underestimated: Forest foods improve the quality and size of daily meals for people around the world. Trees and forests provide a source of energy to ensure that foods are cooked properly. Income from forestry activities helps households purchase the right kinds of foods in sufficient quantities. This video uses a series of slides enhanced by computer graphics to follow the lives of three people in Ghana, Amazonia, and Thailand, illustrating how they benefit nutritionally from the forest. More generally, the video illustrates the many ways trees and forests help alleviate malnutrition and improve food security and emphasizes the need to consider nutrition issues in project designs, management plans, and national development policies so that the contribution of forestry to food security can be strengthened and enhanced.

Fury for the Sound: The Women at Clayoquot
Type: Video
Run time: 86 minutes (52-minute film also available)
Source: Bullfrog Films

A documentary described as "powerful and inspiring," this video shows how a small group of dedicated women, some of them elderly, organized to protest the clear-cutting of the temperate rainforest in Clayoquot Sound off Vancouver Island. The women eventually became part of "the largest civil disobedience action in Canadian history" and hundreds were jailed because they formed a human blockade to prevent logging trucks from entering the rainforest to clear-cut old-growth trees.

Global Warming: Turning Up the Heat
Type: Video (in two parts for classrooms)
Run time: 46 minutes total
Source: Bullfrog Films

For student and adult audiences, this video asks why scientific warnings about climate change due to the greenhouse effect have not been heeded even though there was broad agreement at the international Earth Summit in 1992 that global warming poses a major threat to the Earth. The program shows how human activities have contributed to global warming and examines some of the possible solutions to alleviate the problem.

The Greenhouse Effect
Type: Video
Run time: 20 minutes
Source: Phoenix/BFA/Coronet/MTI

One of nine titles in "The Global Environment Series," this film examines the world's changing climate as it relates to the greenhouse effect. It also explains what steps are being taken to try to alleviate global warming.

Hungry for Profit
Type: Video
Run time: 86 minutes
Source: New Day Films

This documentary looks at the connections between world hunger and global agribusiness. Multinational corporations, argues director Robert Richter, "buy up huge tracts of land in the developing world to produce food for export, displacing local farmers and leaving the host countries less able to feed their own people." The underlying theme: "Agribusiness has contributed to the destruction of tropical rainforests, the expansion of deserts and the huge growth in slums and shantytowns around Third World cities, swollen by the migration of small farmers to the cities in search of work."

**In Good Hands: Culture and Agriculture
in the Lacandon Rain Forest**
Type::   Video
Run time: 27 minutes
Source: The Video Project

Although inefficient agricultural methods are destroying much of the world's rainforests, the Lacandon Maya of Chiapas in southern Mexico have been practicing a sustainable form of rainforest farming for centuries. Dr. James Nations, an ecological anthropologist, worked with the Lacandones for many years, and this film depicts Nations as he interacts with three Lacandon elders who show how they farm in the forests. The video also examines how culture, mythology, and religion influence their agricultural methods. The Lacandon approach to farming provides the basis for a practical alternative to destructive rainforest farming. However, only a few families still practice these ancient methods, as modernization and civil strife threaten their traditional way of life.

Journey into Amazonia
Type: Video (two volumes)
Run time: 3 hours total
Source: PBS Home Video

This program shows nature's "nonstop high-wire act." The videos provide a wild adventure in the Amazon rainforest, where the world's mightiest river shapes and destroys natural habitats during seasonal changes.

Lacandona: The Zapatistas and Rainforest of Chiapas, Mexico
Type: Video
Run time: 27 minutes
Source: The Video Project

Interviewed for this video are national and international activists who are working to help the indigenous Zapatistas in their struggle for self-determination and protection of their ancestral homelands. Zapatista communities are located on land with large oil reserves, natural gas deposits, and timber stands. Wealthy local landowners, the Mexican government, and international investors are among those attempting to take over the land.

The Last of the Hiding Tribes
Type: Video (series of three tapes)

Run time: 50 minutes each
Source: Bullfrog Films

Set in the Brazilian Amazon, this series of three videos was produced by Adrian Cowell, world-renowned documentary storyteller, who traces the history of three tribes near extinction. One part of the series, "Return from Extinction," tells about the recent history of the Panara tribe, who fled into the forest to escape attacking Portuguese in the eighteenth century and were thought to be extinct. "The Fate of the Kidnapper" tells the story of the Uru Eu Wau Wau, and "Fragments of a People" describes the Ava-Canoeiro, descendants of runaway slaves from the once feared Carijo tribe. Each of the three films is presented in two parts for classroom viewing and discussion.

The Living Edens: Manu, Peru's Hidden Rain Forest
Type: Video
Run time: 60 minutes
Source: PBS Home Video

This video is one of a series depicting "Living Edens." Narrated by Edward James Olmos, it presents a virtual field trip to the Manu Biosphere Reserve along the eastern base of the Peruvian Andes. Sustained by the great river Manu, this unspoiled paradise teems with a dizzying kaleidoscope of rare species, and the video includes scenes of colorful parrots and the howler monkey, with its unearthly screech.

Man and the Biosphere
Type: Video (twelve separate tapes)
Run time: 28 minutes each
Source: Films for the Humanities

This series of videos, based on UNESCO's Man and the Biosphere program, shows the work of botanists, biologists, geologists, and demographers who "examine the realities of ecological concerns in the framework of political realities." The videos "illustrate the problems and concerns of preserving life, including human life, on Earth, and demonstrate numerous environmental projects that have successfully met the needs of both humankind and nature. Among the videos are "The Tropical Rain Forest" and "Preserving the Rain Forest."

On the Edge of the Forest
Type: 16mm film

Run time: 32 minutes
Source: Bullfrog Films

Although produced during the 1970s, this program's plea for a commonsense approach to human behavior that will preserve the planet is still relevant today. Economist E. F. Schumacher narrated the film shortly before his death. He walks through a forest in Australia, talking directly to the viewer. He describes the efficiency of the balanced forest ecosystem, explaining the need to preserve this balance rather than ravage the environment in the name of economics and progress.

Our Fragile World
Type: Video (series of four tapes)
Run time: 15 minutes each
Source: Films for the Humanities

Designed to help students understand the interrelationships of plants and animals in ecosystems, this four-part series includes explorations of the rainforest of the Khutzeymateen Valley, just south of the Alaska Panhandle, the coastal temperate rainforest in Clayoquot Sound in British Columbia, and the Costa Rican rainforest.

Rain Forests: Proving Their Worth
Type: Video
Run time: 30 minutes
Source: The Video Project

This video explores the question of whether a living rainforest is of greater economic value than one that has been cut down for timber or other purposes. The international marketing of foods, cosmetics, and crafts derived from tropical forests may provide native inhabitants with the economic means to protect their vanishing lands. Several hopeful developments are presented: new groves of peach palms in Costa Rica, bustling markets on the Amazon, and environmentally conscious American companies that successfully use and market renewable forest products. The video also discusses the many obstacles that remain. A study guide is included.

Secret of the Rain Forest: The Tree Canopy
Type: Video
Run time: 51 minutes
Source: Films for the Humanities

Teams of biologists shown in this film unlock the secret of the rainforest canopy high above the ground. The biologists describe how plants and animals living in the canopy are vital to the rainforest's ecosystem and the forest's survival, and they also explain why the soil of the tropical rainforest contributes little to its lush growth.

Sustainable Environments
Type: Video
Run time: 34 minutes
Source: The Video Project

This program explores all aspects of sustainability, from definitions of terminology to practical and theoretical illustrations of sustainable environments created in nature and by humans. The film shows major changes that people have produced in their environment during the past 100 years, affecting the land, forests, water, and even the atmosphere. Ways to reverse these effects and work toward sustainable solutions are addressed in the film by community planners, energy experts, architects, farmers, horticulturists, and biologists.

Tropical Rain Forest
Type: Videodisc
Run time: 13 minutes
Source: Phoenix/BFA/Coronet/MTI

One of seven titles in "The Biomes Series," this film explores life in the tropical rainforest. It shows the four distinct layers of the forest, with its separate habitats for plants and animals: the emergent layer at the topmost crown of the rainforest, the canopy layer that forms an umbrella over the forest floor, the understory, and the sparsely populated forest floor.

Tropical Rainforest
Type: Videodisc
Run time: 40 minutes
Source: Image Entertainment

Originally filmed in IMAX, this video illustrates the diversity and beauty of life in the forests, featuring sound and film images of birds and primates of the forest canopy and insects of the forest floor. The film also depicts the researchers who attempt to understand these forests and explains how the people of El Salvador are attempting to reclaim their land, water, and forests after devastating wars.

Yanomami: Keepers of the Flame
Type: Video
Run time: 58 minutes
Source::   The Video Project

Television star Michel Dorn hosts this documentary that portrays the journey of a group of journalists, anthropologists, and doctors who traveled to a Yanomami village in the Venezuelan rainforest. Considered the last intact indigenous group in the Americas, the Yanomami continue in a way of life that has changed little over thousands of years. But their existence is being threatened by gold miners, urban developers, and the diseases brought by people of Western cultures. The film includes a brief history of indigenous people in the Americas, an in-depth look at the Yanomami way of life from a Western perspective, and concludes with a plea to preserve Yanomami culture.

Addresses and URLs of Publishers and Distributors

ABC-CLIO
130 Cremona Drive
P.O. Box 1911
Santa Barbara, CA 93116
http://www.abc-clio.com

Abbeville Press
22 Cortlandt Street
New York, NY 10007
http://www.abbeville.com/

Abrams Books
100 Fifth Avenue
New York, NY 10011
http://www.abramsbooks.com/

Academic Press
525 B Street, Suite 1900
San Diego, CA 92101
http://www.apnet.com/

Addison-Wesley
One Jacob Way
Reading, MA 01867
http://www.awl.com/aw/

Anchor/Doubleday
1540 Broadway
New York, NY 10036
http://www.bdd.com

Basic Books
10 E. Fifty-third Street
New York, NY 10022
http://www.basicbooks.com/

Beacon Press
25 Beacon Street
Boston, MA 02108
http://www.beacon.org

Birkhäuser Boston
675 Massachusetts Avenue
Cambridge, MA 02139
http://www.birkhauser.com

Blackstone Audio Books
Box 969
Ashland, OR 97520
http://www.blackstoneaudio.com/

Books on Tape, Inc.
P.O. Box 7900
Newport Beach, CA 92658
http://www.booksontape.com

Bullfrog Films
P.O. Box 149
Oley, PA 19547
http://www.bullfrogfilms.com

California Institue of Public Affairs
P.O. Box 189040
Sacramento, CA 95818
http://www.cipahq.org

Cambridge University Press
40 West Twentieth Street
New York, NY 10011
http://www.cup.org/

Cato Institute
1000 Massachusetts Avenue NW
Washington DC 20001
http://www.cato.org/

Cavendish Publishing Limited
The Glass House, Wharton Street
London WC1X 9PX
United Kingdom
http://www.cavendishpublishing.com

Chronicle Books
85 Second Street, 6th Floor
San Francisco, CA 94105
http://www.chronbooks.com

Columbia University Press
61 West Sixty-second Street
New York, N.Y. 10023
http://www.columbia.edu/cu/cup/

Crown Publishing
1601 McCormick Drive
Largo, MD 20774
http://www.crownbooks.com/

Cultural Survival
215 Prospect Street
Cambridge, MA 02139
http://www.cs.org

Dioscorides Press/Timber Press
133 S.W. 2nd Avenue #450
Portland, OR 97204

Dutton/Penguin/Putnam
345 Hudson Street
New York, NY 10014
http://www.penguinputnam.com

Ecotrust
1200 N.W. Naito Parkway, Suite 470
Portland, OR 97209
http://www.ecotrust.org

Facts on File
11 Penn Plaza
New York, NY 10001
http://www.factsonfile.com

Filmakers Library
124 East Fortieth Street
New York, NY 10016
http://www.filmakers.com/

Films for the Humanities & Sciences
P.O. Box 2053
Princeton, NJ 08543
http://www.films.com

Four Walls Eight Windows Press
39 West Fourteenth Street, Room 503
New York, NY 10011
http://www.fourwallseightwindows.com

Franklin Watts/Grolier
90 Sherman Turnpike
Danbury, CT 06816
http://publishing.grolier.com

W. H. Freeman
41 Madison Avenue
New York, NY 10010
http://www.whfreeman.com

Gibbs Smith Publishers
P.O. Box 667
Layton, UT 84041
http://www.gibbs~smith.com

Harvard University Press
79 Garden Street
Cambridge, MA 02138
http://www.hup.harvard.edu

Harmony Books/Random House
201 East Fiftieth Street
New York, NY 10022
http://www.randomhouse.com/

Houghton Mifflin
222 Berkley Street
Boston, MA 02116
http://www.hmco.com

Humanities Press International
165 First Avenue
Atlantic Highlands, NJ 07716

Image Entertainment
9333 Oso Avenue
Chatsworth, CA 91311
http://www.image-entertainment.com

Island Lake Press
P.O. Box 710
Island Lake, IL 60042

Island Press
1718 Connecticut Avenue NW, Suite 300
Washington, DC 20009
http://www.islandpress.org

Little Brown
1271 Avenue of the Americas
New York, NY 10021
http://www.twbookmark.com/

Living Planet Press
30408 34th Place NW
Washington, DC 20016

Louisiana State University Press
P.O. Box 25053
Baton Rouge, LA 70894
http://lsumvs.sncc.lsu.edu/lsupress/

Lotus Press
P.O. Box 325
Twin Lakes, WI 53181
http://www.lotuspress.com/

The Lyons Press
123 West Eighteenth Street
New York, NY 10011
http://www.lyonspress.com/

Macmillan
1633 Broadway
New York, NY 10019
http://www.mgr.com

McGraw Hill
1221 Avenue of the Americas
New York, NY 10020
http://www.mcgraw-hill.com

Merril Press
12500 N.E. 10th Place
Bellevue, WA 98005
http://www.merrilpress.com

Milkweed Editions
1011 Washington Avenue South, Suite 300
Minneapolis, MN 55415
http://www.milkweed.org/1_0.html

The MIT Press
Five Cambridge Center
Cambridge, MA 02142
http://mitpress.mit.edu/

Multi Dimensional Communications
Orange Cherry Software
P.O. Box 390
69 Westchester Avenue
Pound Ridge, NY 10576
http://www.orangecherry.com

National Geographic
School Publishing
1145 Seventeenth Street NW
Washington, DC 20036
http://www.nationalgeographic.com/education

New Day Films
22-D Hollywood Avenue
Ho-ho-kus, NJ 07423
http://www.newday.com/

North America and Caribbean Regional Center
5400 Grosvenor Lane
Bethesda, MD 20814
http://www-trees.slu.se/publ/video.htm

North Atlantic Books
1456 Fourth Street
Berkeley, CA 94710
http://www.northatlanticbooks.com/

NorthWord Press
5900 Green Oak Drive
Minnetonka, MN 55343

Oregon State University Press
101 Waldo Hall
Corvallis, OR 97331
http://www.orst.edu/dept/press/osupress.htm

Overlook Press
One Overlook Drive
Woodstock, NY 12498
http://www.overlookpress.com/

Oxford University Press
198 Madison Avenue
New York, NY 10016
http://www.oup-usa.org/

Palgrave
Houndmills
Basingstoke,Hampshire RG21 6XS
United Kingdom
http://www.palgrave.com/

PBS Home Video
P.O. Box 751089
Charlotte, NC 28275
http://www.pbs.org/

Phoenix/BFA/Coronet/MTI
2349 Chaffee Drive
St. Louis, MO 63146
http://www.phoenix-bfa-coronet.com

Prentice Hall
Upper Saddle River, NJ 07458
http://vig.prenhall.com/

Princeton University Press
41 William Street
Princeton, NJ 08540
http://pup.princeton.edu/

Rainforest Action Network (RAN)
221 Pine Street, Suite 500
San Francisco, CA 94104
http://www.ran.org/

Random House
1540 Broadway
New York, NY 10036
http://www.randomhouse.com/

Resources for the Future
1616 P Street NW
Washington, DC 20036
http://www.rff.org

Richter Productions
330 West Forty-second Street
New York, NY 10036

Running Wolf Press
P.O. Box 3011
Sequim, WA 98382
http://www.olympus.net/personal/runningwolf/

St. Martin's Press
175 Fifth Avenue
New York, NY 10010
http://www.stmartins.com

Scarecrow Press
15200 NBN Way
Blue Ridge Summit, PA 17214
http://www.scarecrowpress.com

Sierra Club Books
Department WM
85 Second Street
San Francisco, CA 94105

Simon & Schuster
1230 Avenue of the Americas
New York, NY 10020
http://www.simonsays.com

Springer-Verlag
175 Fifth Avenue
New York, NY 10010
http://gonzo.springer-ny.com/

University of California Press
2120 Berkeley Way
Berkeley, CA 94720
http://www.ucpress.edu/

University of Chicago Press
5801 Ellis Avenue
Chicago, IL 60637

University of London
Institute of Latin American Studies
31 Tavistock Square
London WC1H 9HA
United Kingdom
http://www.sas.ac.uk/ilas

University of Washington Press
Marketing Department
P.O. Box 50096
Seattle, WA 98145
http://www.washington.edu/uwpress/

The Video Project
P.O. Box 77188
San Francisco, CA 94107
http://www.videoproject.net

Westview Press
Perseus Books Group
5500 Central Avenue
Boulder, CO 80301
http://www.westviewpress.com

World Wildlife Fund
Avenue du Mont-Blanc CH-1196
Gland, Switzerland
http://www.panda.org/home.cfm

(John) Wiley & Sons
605 Third Avenue
New York, NY 10158
http://www.wiley.com

W. W. Norton & Company
500 Fifth Avenue
New York, NY 10110
http://www.wwnorton.com

Yale University Press
302 Temple Street
New Haven, CT 06520
http://www.yale.edu/yup

Glossary

afforestation   The practice of planting trees to create a forest (see reforestation).

agronomy   The application of plant and soil sciences to crop production.

ancient forest   Often called an old-growth forest—a forest that in general has a large number of old trees that have reached "maturity" and no longer produce new wood each year. Ancient forests are part of many coastal rainforests in the U.S. Pacific Northwest.

atmosphere   The layer of gases that surround the Earth.

biome   An ecological region defined by its climate—the temperature and the amount of precipitation—and by the plants and animals that have adapted to the particular climatic conditions that occur there. There are a variety of biomes, including the desert biome, the rainforest biome, and the tundra biome.

boreal forest   A forest area of the northern temperate zone and Arctic region.

botanist   A person who specializes in the study of plants.

carbon cycle   The circulation of carbon that begins as green plants absorb carbon dioxide from the air. With energy they obtain from the sunlight, plants then combine the carbon dioxide with water to produce glucose, giving off oxygen as a waste product. The oxygen is inhaled by animals and people, who in turn exhale carbon dioxide produced when food is oxidized, or "burned" by their bodies.

carbon A natural element that occurs in all organic compounds and many inorganic compounds.

carbon dioxide (CO_2) One of the colorless gasses that makes up the air and is released through the respiration of living organisms.

clear-cutting The process of cutting down trees in wide patches of forest land.

climatologist A person who specializes in the study of climate.

cloud forest A tropical forest at high elevations often shrouded in clouds.

community In ecology, all the plants and animals in a particular environment.

conifer A tree that bears cones, such as a pine or fir tree.

conservation The wise use of natural resources to reduce waste and loss.

cultivar A plant produced by cultivation.

deciduous tree A tree that sheds or loses its foliage at the end of a season.

desertification The process by which the amount of available soil water decreases or is eliminated because of such human activities as overpumping of wells or using agricultural methods that deplete soil moisture, thereby drying out the soil so that only desert-type plants are able to grow.

developed countries Industrialized nations, or those that are primarily industrial and in which industry provides most of the national income.

developing countries Nations that derive most of their income from agriculture and are also developing industries that produce manufactured goods for export.

ecology The scientific study of living things and their relationships with their environment and with each other.

ecologist A person who specializes in the study of ecology.

ecosystem All organisms living in a particular physical environment, together making up a unit that tends to function as one; changes in any one part of an ecosystem bring about changes in other parts of the system.

ecotourism Travel that promotes environmental awareness and conservation and provides income for local people who have a vested interest in protecting natural resources.

endangered species Species threatened with extinction.

entomologist A person who specializes in a branch of zoology that deals with insects.

environment The entire area—including physical, chemical, biological, and other factors—surrounding a particular organism or object.

epiphyte A plant that uses another plant as a host but does not act as a parasite, receiving its nutrients from the air or trapping food (insects) in its leaves.

erosion The condition in which natural forces such as rain, wind, and gravity remove soil from the surface of the earth.

ethnobotany The study of the relationship between plants and people.

evaporation The process by which a substance changes from a liquid form into a vapor.

evapotranspiration The process by which evaporation and transpiration together release moisture from plants into the atmosphere.

extinction The total elimination of all members of a species of plants or animals.

extractive reserves A section of land set aside for preservation.

fauna The animals of a region.

flora The plants of a region.

forest cover Land area covered with a forest.

fossil fuel Any fuel derived from decayed organisms, such as coal, gas, or oil.

greenhouse effect A common term for the theory that a buildup of gases in the atmosphere leads to global warming. Also known as *global warming*.

Hodgkin's disease A progressive and sometimes fatal disease in which the lymph nodes, spleen, and often the kidneys and liver become inflamed and enlarged. The active ingredient in a treatment for the disease comes from the rosy periwinkle, native to the rainforests of Madagascar.

Gaia hypothesis A theory that the Earth is a self-regulating organism and creates the conditions under which life exists or thrives.

habitat A type of environment suitable for a particular organism and where that organism is likely to live.

holistic approach In biology or ecology, an emphasis on an entire organism and the interrelatedness of its parts, as opposed to a focus on the individual parts that make up the whole.

indigenous Native to a particular environment.

manifesto A proclamation or public declaration of an individual's or a group's intentions.

mammal An animal that belongs to a class in which the female have milk-producing glands.

Nongovernmental organization (NGO) Usually an international aid or advocacy group funded by private donations and not part of any government.

nonrenewable natural resource A natural resource that cannot be replaced in its original state.

nurse log A decaying log that provides nutrients for seedlings that have sprouted on its surface.

nutrient A substance that an organism needs for normal growth and development.

nutrient cycling The constant reuse of nutrient molecules by plants and animals.

old-growth forest *See* ancient forest.

peasant Someone who is part of an agricultural labor force and earns only subsistence income. Most peasants own no land.

photosynthesis The process that green plants use to absorb sunlight for energy and produce food from carbon dioxide and water.

reforestation Replanting a cut or burned forest or preparing deforested land for tree planting.

sediment Particles of soil and other solid matter that move from their original site to settle on a land surface or at the bottom of a waterway or body of water.

seedling A young tree, usually less than five years old.

seed tree A tree left during a forest cutting in order to provide seeds to regenerate the forest.

shifting cultivation The practice of farming land until the soil degrades then moving on to more fertile plots and continuing the practice for survival.

silt Sedimentary material containing small rock particles.

silviculture The care and cultivation of forests.

slash-and-burn agriculture A farming method in which trees are cut and burned to clear land for growing crops.

species A category used in biology to classify organisms that consist of a group of individuals that are capable of interbreeding and producing fertile offspring.

subspecies A classification for plants or animals that ranks immediately below species.

sustainable development An economic system that manages natural resources to meet present human needs while maintaining those resources to meet future human needs.

sustained yield A harvesting method that reaps nonrenewable resources at a rate that can be maintained long into the future.

swidden agriculture The use of ancient slash-and-burn practices to clear land for farming.

Taxus baccata A chemical found in the bark of yew trees and used in the treatment of cancer.

technology The tools and techniques used to accomplish a purpose, such as to make products or to apply science.

threatened species Any species that is likely to become endangered.

transpiration The process of giving off vapor.

wilderness area Lands designated by the U.S. Congress or a federal agency as wilderness and set aside for study.

wildlife refuge A protected wildlife habitat that may be maintained for such recreational activities as fishing and hunting.

zoology A branch of biology that deals with the study of animals and animal life.

Index

DISCARDED
New Hanover County Public Library from

Kathlyn Gay is the author of more than 100 books, including young adult books, encyclopedias, teacher manuals, and portions of textbooks. Most of her work focuses on social and environmental issues, culture, history, and communication. In 1983, her book *Acid Rain* was selected as an "Outstanding Book" by the National Council for Social Studies and National Science Teachers' Association. *Silent Killers* received the same award in 1988. *Global Garbage: International Trade in Toxic Waste* was chosen as a Notable Book for Young People in 1993. Some of her publications issued since 2000 include such young adult titles as: *Eating Disorders; Epilepsy; Leaving Cuba: From Pedro Pan to Elian;* and *Silent Death: The Threat of Biological and Chemical Terrorism.*